CHAOS AND THE CHANGING NATURE OF SCIENCE AND MEDICINE: AN INTRODUCTION

CHAOS AND THE CHANGING NATURE OF SCIENCE AND MEDICINE: AN INTRODUCTION

Mobile, AL   April 1995

EDITOR
Donald E. Herbert
University of South Alabama

ASSOCIATE EDITORS
Paul Croft
Daniel S. Silver
Susan G. Williams
Marie Woodall

American Institute of Physics

AIP CONFERENCE
PROCEEDINGS 376

Woodbury, New York

Authorization to photocopy items for internal or personal use, beyond the free copying permitted under the 1978 U.S. Copyright Law (see statement below), is granted by the American Institute of Physics for users registered with the Copyright Clearance Center (CCC) Transactional Reporting Service, provided that the base fee of $6.00 per copy is paid directly to CCC, 222 Rosewood Drive, Danvers, MA 01923. For those organizations that have been granted a photocopy license by CCC, a separate system of payment has been arranged. The fee code for users of the Transactional Reporting Service is: 1-56396-442-2/ 96 /$6.00.

© 1996 American Institute of Physics

Individual readers of this volume and nonprofit libraries, acting for them, are permitted to make fair use of the material in it, such as copying an article for use in teaching or research. Permission is granted to quote from this volume in scientific work with the customary acknowledgment of the source. To reprint a figure, table, or other excerpt requires the consent of one of the original authors and notification to AIP. Republication or systematic or multiple reproduction of any material in this volume is permitted only under license from AIP. Address inquiries to Office of Rights and Permissions, 500 Sunnyside Boulevard, Woodbury, NY 11797-2999; phone 516-576-2268; fax: 516-576-2499; e-mail: rights@aip.org.

L.C. Catalog Card No. 96-85220
ISBN 1-56396-442-2
DOE CONF- 9504196

Printed in the United States of America

CONTENTS

Preface vii
Steering Committee ix
Corporate Support ix
Institutional Support ix
Night Thoughts of a Classical Physicist xi

Overview of Nonlinear Dynamical Systems and Complexity Theory 1
    D. E. Herbert
Catastrophe Theory: What It Is—Why It Exists—How It Works 35
    R. Gilmore
Fractal Structures and Processes 54
    J. B. Bassingthwaighte, D. A. Beard, D. B. Percival, and G. M. Raymond
Chaos, Dynamical Structure, and Climate Variability 80
    H. B. Stewart
Coherent Structures Amidst Chaos: Solitons, Fronts, and Vortices 115
    D. K. Campbell
From Self-Organization to Emergence: Aesthetic Implications
of Shifting Ideas of Organization 133
    N. K. Hayles
Two Examples of Chaotic Dynamics in Fluids 158
    M. Gorman
Applications of Chaos in Biology and Medicine 175
    W. L. Ditto

Author Index 203

PREFACE

These Proceedings are the expanded versions of the lectures delivered at the one-day Workshop "Introduction to Chaos and the Changing Nature of Science and Medicine" that was presented at the Adam's Mark Hotel in Mobile, Alabama on 29 April 1995, under the joint sponsorship of the University of South Alabama and the American Association of Physicists in Medicine.

The interest in and support of the Workshop was sparked by the lecture "Chaos Theory" given by Professor James Yorke of the University of Maryland in the University of South Alabama Forum series of 1992. These permitted us to invite seven of the most outstanding workers and writers in the field of nonlinear dynamics to deliver lectures at the Workshop and to provide copies of their lectures for publication by the American Institute of Physics: James Bassingthwaighte, M.D., Ph.D., University of Washington; David Campbell, Ph.D., University of Illinois, Urbana-Champaign; William Ditto, Ph.D., Georgia Institute of Technology; Robert Gilmore, Ph.D., Drexel University; Michael Gorman, Ph.D., University of Houston; N. Katherine Hayles, Ph.D., University of California Los Angeles; H. Bruce Stewart, Ph.D., Brookhaven National Laboratory.

The Workshop was directed to providing those attending with a survey of the principal manifestations and consequences of the ubiquitous presence of nonlinearity in Nature—catastrophes, fractals, low-dimensional deterministic chaos, and nonlinear waves and vortices—together with several of the most striking and useful of their recent applications.

The success of the enterprise required the well-informed and dedicated work of a small number of people. The creative practical suggestions and sophisticated counsel and direction provided by Susan G. Williams, Ph.D., Daniel S. Silver, Ph.D., and Paul Croft, Ph.D. of the Steering Committee were crucial to every aspect of the success of the Workshop from its inception to publication of the Proceedings. Special thanks are due to Marie Woodall, B.A., Secretary to the Steering Committee, for her highly competent, imaginative, and dedicated assistance in every aspect of the preparation and presentation of the Workshop: communication, organization, supervision, editing, processing, and proofreading. Special thanks are also due to Ray Butler for his excellent audio-visual support of the lectures and to Frank Vogtner for his skills and talents in preparation of many of the graphs and tables. Finally, we wish to thank Michael Hennelly and Anita Hargrove of the American Institute of Physics for their considerable help and encouragement in the preparation of these lectures for publication.

The Workshop was generously supported by the University of South Alabama:

College of Medicine, Sam Strada, Ph.D., Senior Associate Dean
    Dept. of Radiology, Steven Teplick, M.D., Chair
    Dept. of Physiology, Aubrey Taylor, Ph.D., Chair
College of Arts and Sciences, Lawrence Allen, Ph.D., Dean
    Dept. of Marine Sciences, Robert Shipp, Ph.D., Chair
    Dept. of Philosophy, Harold Baldwin, Ph.D., Chair
    Dept. of Physics, S. L. Varghese, Ph.D., Chair
College of Engineering, David Hayhurst, Ph.D., Dean
College of Education, George Uhlig, Ph.D., Dean
Office of Graduate Studies and Research, James Wolfe, Ph.D., Dean

The Workshop also had strong support from corporate friends of the University and AAPM: **Dupont Corp.** (Louis Eaton), **Gammex RMI Corp.** (Charles Lescrenier, D.Sc.), **IBM Corp.** (Steve Boecler), **Konica Corp.** (David Vandal), **Oldelft Corp.** (Stanley Babinski), **Siemens Medical Systems** (Russell Porter), and **Varian Medical Systems** (Thomas Rhea, Ph.D.).

It is hoped that these Proceedings will help to sustain and further inform the interest in nonlinear dynamics that was abundantly evident at the Workshop.

Donald Herbert, Ph.D.
Chairman/Editor

Steering Committee

Donald Herbert, Ph.D. (Chairman)
Department of Radiology
University of South Alabama

Daniel Silver, Ph.D.
Department of Mathematics
and Statistics
University of South Alabama

Marie Woodall, B.A. (Secretary)
Department of Radiology
University of South Alabama

Paul Croft, Ph.D.
Department of Geology
and Geography[*]
University of South Alabama

Susan Williams, Ph.D.
Department of Mathematics
and Statistics
University of South Alabama

[*] Presently at Jackson State
University, Dept. of Physics
and Atmospheric Sciences

Corporate Support

Dupont Corporation
IBM Corporation
Oldelft Corporation
Varian Incorporated

Gammex Corporation
Konica Incorporated
Siemens Incorporated

Institutional Support

College of Arts and Sciences: Dean's Office, Department of Marine Sciences, Department of Philosophy, and Department of Physics

College of Education: Dean's Office

College of Engineering: Dean's Office

College of Medicine: Dean's Office, Department of Physiology, and Department of Radiology

Office of Graduate Studies and Research

Sponsored by The University of South Alabama and The American Association of Physicists in Medicine

The Geheimrath brought up the complexity of phenomena, which experimentalists understood better than theorists because of their practical experience. Jakob pointed out that the history of physics showed that connections in nature were less complex than previously imagined."

...

"... physicists must give up the assumption that underlay the previous foundations of theoretical physics which was that only continuous processes could occur in nature."

Night Thoughts of a Classical Physicist

Overview of Nonlinear Dynamical Systems and Complexity Theory

Donald E. Herbert, Ph.D.

Department of Radiology, College of Medicine
University of South Alabama, Mobile, AL 36688

Abstract: *A brief overview is presented of the principal elements of "nonlinear dynamics": catastrophes, fractals, chaos, solitary waves, and coherent and dissipative structures. The text is followed by a set of 10 portraits of the strange and violent world of nonlinear dynamics.*

INTRODUCTION

"The style of any mathematic which comes into being, then, depends wholly on the culture in which it is rooted, the sort of mankind it is that ponders it" (1).

Oswald Spengler, 1918
The Decline of the West

Spengler's thesis that mathematics is a "culture clue" recently resurfaced in the writing of the historian Eric Hobsbawm (*The Age of Extremes*, 1994): "Again, the lay observer cannot but notice the emergence, within the field of thought remotest from flesh-and-blood human life, of two mathematical sub-fields known respectively as 'catastrophe theory' ... and 'chaos theory' ... . The one ... claimed to investigate the situations when gradual change produced sudden rupture ... the other ... modelled the uncertainty and unpredictability of situations in which apparently tiny events ... could be shown to lead to huge results elsewhere ... Those who lived through the later decades of the [20th] century had no difficulty in understanding why such images as chaos and catastrophe should come into the minds of scientists and mathematicians also" (2). See also N.K. Hayles, 1989 (3).

The study of catastrophe and chaos, together with fractal geometry and nonlinear waves, arose in the 1960s and early 1970s (around the beginning of the historical epoch called *The Landslide* by Hobsbawm and coeval with the publication of the then-heretical *The Structure of Scientific Revolutions* by T. Kuhn (4).) All but catastrophe theory, which was invented by R. Thom in the 1970s, had distinguished 19th and early 20th century precursors. The "style of mathematic" is <u>nonlinear dynamics</u>. In this overview we present a brief summary of its elements.

What is nonlinear dynamics? A dynamical system is any <u>system</u>, biological, chemical, physical, social, etc., that <u>evolves over time</u>. More particularly, if x is a set of variables describing the instantaneous <u>state</u> of a system, e.g., position and velocity of a mass, concentration of a chemical, etc., then the evolution over time of a dynamical system can be described by a set of ordinary differential equations in x in which the rate of change of any variable of the set at a given time is a function of the values of all the variables of the set. There are two kinds of dynamical systems; those described by linear equations and those described by nonlinear equations. For linear systems the rate function includes only variables to the first power. For nonlinear systems the rate function includes, in addition, nonlinear terms: powers, cross-products, transcendental terms such as trigonometric functions, and other exotica.

Nonlinear dynamical systems have an incredibly rich and complex spectrum of recurrent behaviours — equilibria, periodic, quasiperiodic, and aperiodic (chaotic) — and modes of change of behaviour (bifurcations) — continuous, discontinuous, and explosive — well beyond anything to be seen in linear dynamical systems. Indeed, to paraphrase the remarks of the biologist J.B.S. Haldane concerning the Universe, the behaviour of nonlinear dynamical systems is not only stranger than we suppose, it is even stranger than we can suppose. Complexity resides in nonlinear systems.

CATASTROPHES

Many of the most interesting and important natural phenomena involve sudden jumps and discontinuities, yet the vast majority of available mathematical techniques were designed for the quantitative study of continuous behaviour (G. Leibniz, co-inventor of the calculus in the 17th century, hopefully asserted that "Nature does not make jumps" (5).) Catastrophe theory provides a natural continuous description of discontinuous behaviour. It is the proper study of <u>qualitative change</u>. A catastrophe is a discontinuous change in the behaviour, or structure, of nonlinear dynamical systems that occurs as one or more system parameters, e.g., temperature, frequency, etc., is varied continuously. Processes in which an infinitesimal perturbation may cause very large changes in the outcome are typically catastrophes (Poincaré's, "... observable effect without an observable cause")[1]. Familiar examples of catastrophes are the equilibrium phase transitions, e.g., the discontinuous change in density of water as the temperature changes through the boiling point (first-order phase transition) or the continuous change in the magnetic properties of iron (para- to ferromagnetic) as the temperature changes through the Curie point (second-order phase transition). There is also a change in the crystal symmetry group at the Curie point: cubic to tetragonal (6). Still other examples are the sudden hysteretic discontinuities in amplitude and phase of the oscillation of driven passive nonlinear oscillators as the driving frequency is varied through the oscillator's low-amplitude natural frequency (7). And catastrophes precede the mode-locking, or entrainment (8), of both frequency and phase in either forced or coupled self-excited (auto-catalytic) oscillators (9). We note that entrainment and "jumps" are among the most ubiquitous non-linear phenomena, being observed in all areas of science and engineering. (Entrainment is a crucial concept in understanding the evolution of social, as well as biological, aggregative and cooperative structures.)

The change in symmetry of iron at the Curie point is a typical example of a symmetry-breaking transformation in thermodynamic equilibrium - it is a generalized catastrophe (10). In nonlinear systems in far-from-equilibrium conditions such generalized catastrophes also occur. For example, in a film of liquid in which the effect of a temperature gradient normal to the surface opposes that of the gravitational field, a stationary set of convection cells will abruptly appear as the gradient increases through a critical value (the so-called Benard instability) (11). In this example, the system is spatially uniform before the catastrophe and spatially periodic afterward; it is a non-equilibrium phase transition (a form of crystallization under long-range correlations rather than short-range interactions). The set of convection cells is one example of a new dynamic phase of matter, the <u>dissipative structure</u>, that arises spontaneously in nonlinear systems in non-equilibrium conditions in a process of self-organization (12).

<u>Self-organization</u> - the <u>emergence</u> of complex macroscopic space/time "patterns" from a uniform system - is an emerging paradigm of science in which nonlinear processes in non-equilibrium conditions play

---

[1] Some current sociological manifestations of this defining characteristic of nonlinear systems are nicely limned in R. Frank and P. Cook's *The Winner Take All Society*. The Free Press. NY. 1995

significant roles (13). Symmetry-breaking and entrainment both play crucial roles in the self-organization process in nonlinear systems. Other examples are the vortices in flowing fluids and bistable systems such as membranes and the single-mode laser; in the latter the emitted radiation is incoherent (homogenous) below an input threshold and becomes coherent (inhomogeneous) above it - a second order phase transition. (The laser also provides a vivid instance of phase-entrainment (11).) Still other examples are the self-excited or autocatalytic electronic oscillators that are in a quiescent state below a threshold level of a parameter and in an oscillatory state above it (14). The appearance of periodicity in either space or time is a symmetry-breaking transformation. Symmetry-breaking in systems in thermodynamic equilibrium that describe changes of phase and in systems far from equilibrium leading to dissipative structures are analogous in that in each case it describes the _emergence_ of inhomogeneity or complexity from homogeneity or simplicity (15). (These phenomena provide evidence of Herbert Spencer's thesis of the "instability of the homogeneous".)

The surface of the cusp catastrophe function (16) – an _attractor_ of the nonlinear dynamics - describes a nonlinear wave (e.g., shock waves and solitons) in which the velocity is a function of amplitude as well as a non-linear oscillator in which the resonant frequency is a function of amplitude.

The cusp catastrophe _potential_ function also provides the argument of the density functions of statistical Catastrophe Theory (17, 18), which is an appropriate model for the analysis of data exhibiting bimodal density functions when the usual assumption that such a density function is a mixture of unimodal density functions is inappropriate, e.g., for a homogeneous bistable (and hence nonlinear) system, as in perceptual bistability (19).

Catastrophe theory was initially developed not as a part of theoretical physics as are most branches of applied mathematics, but of theoretical biology in an examination of the fundamental problem of succession of form, i.e., of _morphogenesis_; it describes "the morphology of a process" (20).

FRACTALS

According to Galileo (1610) the language of the universe is written in "triangles, circles, and other geometric figures" (21). But Mandelbrot (1977) noted that this is not always the case: "Clouds are not spheres, mountains are not cones, coast-lines are not circles"; rather, they are fractals (22). "A fractal is a shape made up of parts similar to the whole in some way" (22). That is, a fractal structure is _self-similar_, and it has noninteger dimension. As H. E. Stanley has remarked, the recognition of self-similarity as a hidden symmetry in a wide variety of apparently random and disordered - "messy" - structures that occur in Nature is a remarkable achievement; it describes the "morphology of the amorphous" (22).

Dilation and translation are two of the fundamental symmetries of nature. Fractal structures are invariant under combined translation-dilation operations that comprise the <u>scaling symmetry</u> under which the form is the same no matter on what scale the object is examined (23). Again, Stanley notes that together with the periodic symmetries that describe the spatial rotations and translations as manifested in molecules and crystals, the scaling symmetry is now one of the most important symmetries in both the natural and the biological sciences. But, as M. Schroeder has acutely observed, "Self-similarity is akin to periodicity on a logarithmic scale." (It is of interest that the wavelet transform (24), now beginning to be competitive to Fourier transforms for data compression in the transmission and archiving of medical images, is based upon, "the two essential operations: <u>translation</u> and <u>dilation</u>.")

Mathematically, the periodic symmetries comprise a group, while the scaling symmetries comprise an iterated function system (IFS) (25). The fractal structure is the <u>attractor</u> of the IFS. The middle-thirds Cantor set provides the fundamental example of a deterministic fractal structure invariant under combined translation-dilation symmetries and of the simplest IFS. A Cantor set is a set of points on a curve such that between any two points there are other points of the set and also gaps of finite width. The middle- thirds Cantor set is a fractal set of total length zero and dimension 0.63. Ramified objects whose apparent density or length changes with the length of the measuring scale, are fractals. The respiratory and vascular systems are examples of a random fractal structure invariant under the combined translation-dilation symmetries (26). The fractal morphology characteristic of the growth of aggregates and ramified clusters (e.g., a diffusion-limited aggregate with a fractal dimension of about 1.7) is usually a consequence of a non-equilibrium condition (27, 28). The Lichtenberg figures (fractal dimension $\sim$ 1.7) that arise in dielectric breakdown are described as <u>generalized catastrophes</u> by Thom, i.e., they represent "... the destruction of a symmetry or uniformity" (29). The symmetry of the system changes from uniform translation (homogeneous) to translation-dilation (fractal).

There are <u>fractal processes</u> as well as fractal structures. Such processes often have "1/f" power spectra, i.e., the power is greater at lower frequencies. Examples are fluctuations in the period of the normal heart, in the annual flood levels of river systems, in musical melodies, and in all electronic components. A fractal dimension can be assigned to such processes. For example, the so-called "pink noise", with a $1/f$ power spectrum, has a fractal dimension of about 1.7, the same as a diffusion-limited aggregate; "black noise", with a $1/f^3$ power spectrum, has a fractal dimension of about 1.2, the same as the coast-line of Britain and the cosmological large-scale structure (30). As the exponent of f increases the degree of correlation in the process increases: "white noise" $1/f^0$, fractal dimension 2, is uncorrelated; "black noise" is highly correlated (31, 32).

After the introduction of fractal geometry by Mandelbrot, a key issue is to understand why (and how) fractal structures in space and fractal processes in time arise and the possible relation(s) between them. Self-organized Criticality (33) provides a coherent explanation of the widespread natural occurrence of both <u>fractal structures</u> and <u>fractal processes</u> and their relationship: a nonlinear dynamical system with extended degrees of freedom evolves naturally to the <u>self-organized critical state</u>, independently of initial conditions, producing "$1/f$" noise and a fractal architecture in evolving to a <u>meta-stable</u> final state. The critical state is an <u>attractor</u> of the dynamics. Self-organized criticality has been invoked as an explanation of a variety of catastrophic phenomena from avalanches to earthquakes (34). (Nonlinear oscillators just below the transitions from periodic to aperiodic behaviour or quasi-periodic to periodic behaviour (entrainment) are also sources of "$1/f$" noise.)

CHAOS

O. Rossler has noted that, "If oscillation is <u>the</u> typical behaviour of two-dimensional dynamical systems, then chaos, in the same way, characterizes three-dimensional continuous systems." Deterministic chaos is random-appearing behaviour that resides in the internal (intrinsic) nonlinearity of a deterministic nonlinear dynamical system of low dimension rather than in additive noise (35). (If the efficient cause of deterministic chaos is taken to be nonlinearity, the <u>final</u> cause of chaos in many biological systems may be to prevent <u>entrainment</u> - chaos may have survival value for individuals and species.) A <u>random sequence</u> of events $(x_1, x_2, ..., x_k)$ is one in which any one of several possible events can occur next, each with a specified probability of occurrence, $p(x_i)$, independent of the previous occurrence of any other event. (However, as noted above, in "$1/f$" random sequences, there may be a degree of correlation in some random sequences.) A <u>deterministic sequence</u> of events $(x_1 ..., x_k)$ is one in which only the event that is uniquely related to one or more of the preceding events by a precise law can occur next. For continuous time the law typically has the form of an evolution equation - a "prescription for the

future" (J. Ottino, 1992):
$d\underline{x}/dt = f(\underline{x}; \underline{\alpha})$ autonomous, or $d\underline{x}/dt = g(\underline{x}, t; \underline{\beta})$ non-autonomous where $\underline{x}$ is a set of state variables and $\underline{\alpha}$, $\underline{\beta}$ are sets of adjustable control parameters, e.g., temperature, frequency. There are corresponding equations for discrete time. In a <u>chaotic sequence</u> each event is uniquely determined by the preceding one(s); although it is completely aperiodic it is completely determinate (36).

The nature of the solutions of the evolution equations is a function of the parameters $\underline{\alpha}$, $\underline{\beta}$. The recurrent (and hence bounded) behaviour of a nonlinear dissipative dynamical system for a specified set of values of the parameters ($\underline{\alpha}$ or $\underline{\beta}$) is determined by the point set in phase space to which the trajectory tends asymptotically, i.e., the <u>attractor</u>, and the initial conditions on $\underline{x}$. The phase space for a dynamical system is defined by the set of variables required to specify the dynamical state of the system. For the simple mechanical oscillator these are the position and velocity. A point in phase space identifies the dynamical state of a system at a given instant of time. Convergence of trajectories to an attractor is an example of self-organization in a nonlinear system. The exhaustive set of <u>initial conditions</u> from which trajectories asymptotically approach a given attractor is its <u>basin of attraction</u>. The boundary of the latter is the <u>basin boundary</u>. The recurrent behaviours of principal interest and their respective <u>attractors</u> are: equilibrium (point), periodic (limit cycle and Mobius band), quasi-periodic (torus), chaotic (fractal or strange) (37). Trajectories on a strange attractor diverge exponentially with the maximum separation limited to the finite size of the attractor; the rate of divergence being determined by the largest Lyapunov exponent. Basin boundaries, as well as attractors, may be either smooth or fractal, and their nature as well as size depends on the system parameters. The behaviour of nonlinear dissipative dynamical systems changes qualitatively as the system parameters are varied, e.g., from quasiperiodic to chaotic. These qualitative changes are described as <u>bifurcations</u>. For example, in a self-excited electronic oscillator the symmetry-breaking change from a quiescent state (point attractor) to an oscillatory state (limit cycle attractor) proceeds via a Hopf bifurcation. The symmetry-breaking transition in the Benard instability proceeds via a pitchfork bifurcation (which is a section of the cusp catastrophe function). The transition from incoherent to coherent emission in the laser occurs via a transcritical bifurcation (38). Bifurcations can be described as continuous (subtle or super-critical) or discontinuous (catastrophic or subcritical). The latter can be further described as either <u>dangerous</u> (an attractor appears or disappears and the system jumps to a remote disconnected attractor in a hysteretic discontinuity) or <u>explosive</u> (the attractor explodes to a new, larger attractor that includes the old attractor as a subset) (39). Basin boundaries, as well as attractors, can undergo catastrophic changes in nature, position, and extent as the system parameter changes. In particular, basin boundaries can suddenly jump in position and change from smooth to fractal, a phenomenon described as <u>metamorphosis</u> (40). If the boundary between two <u>non</u>chaotic attractors (e.g., points or limit cycles) is fractal, long-term prediction from initial conditions near the boundary becomes essentially impossible because the final state of the system is exquisitely sensitive to the initial state (41).

The evolution of any real nonlinear system is jointly determined by both probabilistic and deterministic elements, i.e., by "chance and necessity." Between bifurcations the system is deterministic with internal fluctuations small and decaying exponentially. At bifurcations, fluctuations become large and extend over large distances, i.e., there are long-range correlations; decay follows a power-law, and chance determines the branch that the system subsequently follows giving rise to "Order through fluctuations" (Prigogine, 1978). A given sequence of bifurcations may thus lead to a variety of different final states from a given set of initial conditions because at each bifurcation the subsequent path is determined by chance thus introducing into physical and chemical phenomena a historical element that has hitherto been reserved for biological, cultural, and social phenomena.

In the vicinity of a bifurcation the system becomes very sensitive to external as well as internal perturbations. The origins of the <u>asymmetry</u> that is so common in nature, especially in biology (one-handedness of proteins, morphological asymmetries between left and right, etc.) may be explained by the sensitivity of the system in the vicinity of a symmetry-breaking bifurcation to the effects of an external chiral or polar field not explicitly involved in the bifurcation. The perturbed system captures the external asymmetry and builds patterns of preferred chirality or polarity (42).

It is possible, indeed it is common, that at fixed values of the parameters several attractors may co-exist in which case the asymptotic behaviour of the system, e.g., periodic or chaotic, depends on the initial conditions. And although it is widely known that nonlinear systems can have two or more stable states co-existing at a given set of values of the control parameters, the incredible richness of the behaviour spectrum in nonlinear systems is not often appreciated. For example, some experiments suggest that a nonlinear system may have an exceedingly large number of different stable states, some periodic, some quasi-periodic, and some chaotic. Each of these states corresponds to a phase space attractor with its own basin of attraction. There is no systematic way to determine if all basins of attraction have been discovered, even at specified values of the control parameters. "In fact, two independent investigators working on the same kind of system at the same control parameters could observe quite different phenomena because of different ... histories," i.e., <u>different</u> sequences of bifurcations (43).

Deterministic chaos is characterized by short-term predictability and long-term unpredictability arising from an exquisite sensitivity to initial conditions, a feature that distinguishes motion on a strange attractor from that on a noisy limit cycle. Chaotic behaviour includes an infinite number of unstable periodic motions. The first feature can be exploited to <u>identify</u>, the second to <u>control</u> chaotic behaviour. For a chaotic time-series the accuracy of a forecast decays with increasing forecast horizon at a rate given by the Lyapunov exponent, whereas for a stochastic time-series the forecasting accuracy is approximately independent of the horizon. The method of Sugihara and May exploits this distinction to discriminate between "chaotic" and "stochastic" time-series from short runs of real data (44). Ott, Grebogi, and Yorke (OGY) demonstrated that the behaviour of a nonlinear dynamical system can be changed from chaotic to periodic by adjusting an accessible system parameter to stabilize one of the many unstable periodic trajectories that are dense on a chaotic attractor (45). Ditto, Rauseo and Spano implemented the OGY method to achieve "control of chaos" in a parametrically driven physical system (46). Broad band noise in power spectra can arise from either stochastic or deterministic processes, but the decay in spectral power at high frequencies is different in the two cases. For stochastic (noise) processes the spectrum falls off as a power-law, e.g., $1/f$ noise; for low dimensional deterministic chaos processes the spectrum falls off as an exponential. It is maintained that this difference can be used to discriminate between the two processes in some circumstances (47).

A spectacular effect of deterministic chaos in dissipative systems appears to be the irregular occurrence of the El Niño events, characterized by abrupt changes in the direction and velocity of the trade winds, which dramatically affect the Earth's climate on a global scale over a period of several years (48). The El Niño events have profound ecological, economic, and social effects. The chaotic behaviour of these events appears to result from mode-jumping of the system between co-existing overlapping nonlinear resonances. These resonances result from mode-locking of the equatorial Pacific ocean atmosphere oscillator at integral multiples of the earth's seasonal cycle (49, 50).

A less spectacular effect is the suggestion of the presence of a chaotic component in the normal heart rhythms. Different individuals hold different opinions on the latter effect (51).

Chaotic behaviour also occurs in conservative systems although attractors are not, of course, possible in such systems; the phase-space of most Hamiltonian systems is divided so that trajectories with the same energy can be either chaotic or periodic depending on initial conditions (52). H. Poincaré (1896) first discovered this chaotic behaviour in the famous three-body problem (e.g., Earth-Moon-Sun). Subsequent studies of chaos in the (conservative) dynamics of the Solar System have been quite fruitful. The perdurable question of the origin of the Kirkwood gaps in the asteroid belts between Mars and Jupiter has been solved by showing that these are consistent with orbits that display random, infrequent, abrupt, and large excursions in eccentricity, putting them in "crossing situations" with Mars and Earth by which they are removed (in collisions that, in the case of the latter, may have produced the celebrated recurrent "extinctions" of life) (53, 54). It has been generally assumed that the Solar System is quasi-periodic and therefore stable. However, there is now good evidence that the motions of all nine planets of the Solar System are chaotic with an exponential divergence time of only about 5 million years (Lyapunov exponent of $1/5$ myr^{-1}) (55). Thus, it is not possible to exclude the possibility that the orbit of the Earth may abruptly display wild excursions in eccentricity similar to that of the asteroids (a generalized catastrophe whose possibility was remarked by Thom in 1972). Chaotic behaviours in the orbits of the planets of the solar system and in the Pacific ocean thermoclines of the earth both appear to reside in the overlaps of the several resonances of each nonlinear dynamical system.

NONLINEAR WAVES AND VORTICES

In a _linear_ wave pulse the component Fourier harmonics _do not interact_ with each other and, therefore, propagate with independent velocities; consequently, in a dispersive medium the pulse quickly spreads. But in a _nonlinear_ wave pulse the Fourier harmonics interact and under certain conditions acquire an ability to propagate with equal velocities, i.e., coherently. In this case the wave pulse does not disperse; it survives as a pulse for a much longer time than does a linear wave pulse (56). Such a long-lived nonlinear wave packet is a _coherent structure_. (A corresponding interaction - entrainment - of nonlinear oscillators gives rise to _dissipative structures_. Coherent structures and dissipative structures are examples of self-organization in nonlinear systems.) Nonlinearity in the excited medium tends to increase phase speed with wave amplitude, then waves with greater amplitude tend to overtake lower amplitude waves. The result is to make the wavefront steeper over time and distance leading to the formation of _a shock wave_. (A shock wave is a form of the cusp catastrophe (57). In a nonlinear wave the amplitude and velocity interact; in a nonlinear oscillator the amplitude and frequency interact.) In a dispersive medium there are two competing tendencies: a steepening due to nonlinearity and a spreading due to dispersion. If these effects are balanced, a solitary wave or soliton is formed in which the amplitude and half-width are functions of the velocity, i.e., the faster waves are higher and sharper; a dramatic example is the tsunami that drowned more than 100,000 Japanese in 1702 (58). A more familiar example is the various "biological" solitons that transport energy and information within a living system (e.g., the nerve impulse). The solitary waves are self-resonant waves that are mutually transparent in collision (i.e., elastic collisions). Solitons emerge after interacting nonlinearly without change of shape, speed or amplitude. The only effect of a soliton-soliton interaction is a finite, spatial phase shift of the post-interaction trajectory with respect to the pre-interaction trajectory (59). The solitary wave is an example of a coherent structure, i.e., a stable spatial-temporal structure that is formed in nonlinear systems as a result of the concomitant presence of two phenomena with opposite effects such as nonlinearity and dispersion (solitary wave) or nonlinearity and diffusion (shock wave). In a nonlinear dispersive lattice (e.g., a set of coupled nonlinear oscillators) with periodic boundary conditions excitation in a single mode will lead to the formation of solitons (60), which subsequently recombine to recreate the initial condition - the so-called FPU recurrence (61, 62). In this phenomenon the initial excitation energy is not partitioned equally (and irreversibly) among all lattice modes as prescribed by the equipartition theorem, but rather in only a few from which all of it is subsequently recaptured by the initial mode. Unlike the chaotic behaviour of the nonlinear oscillator the nonlinear lattice retains a "memory" of the initial conditions that it eventually recreates. Nonlinear waves (Kelvin waves and Rossby solitons in the Pacific Ocean thermocline) play a role in the ignition and extinction of the El Niño events (49).

Another kind of coherent structure is the anticyclonic vortex, the most dramatic example of which is the Great Red Spot of Jupiter, an anticyclonic vortex (a solitary Rossby vortex) with a diameter much larger than that of the earth that has been stable for over 300 years (63). In anticyclones dispersive spreading is compensated by the nonlinearity, whereas in cyclones such a compensation is absent; therefore, the anticyclones manifest longevity while the cyclones are short-lived structures. The anticyclonic vortices are coherent structures similar to solitons, but the former collide inelastically while the latter collide elastically (64). Both are examples of emergence - self-organization - in nonlinear systems.

CONCLUSION

Since the time of Newton the fundamental properties of physical systems have been explained using differential equations of second order in time to achieve the ultimate goal of science: "full knowledge of nature and thereby complete predictability, if not domination, of its course" (65). The phenomenon of low-dimensional deterministic chaos suggests that we are somewhat further from that goal than was generally thought to be the case before the 1960s. Sir James Lighthill (66) observed at a 1986 international meeting of scientists that "We are all deeply conscious today that the enthusiasm of our forebears for the marvelous achievements of Newtonian mechanics led them to make generalizations in this general area of predictability which, indeed, we may have generally tended to believe before 1960, but which we now recognize were false. We collectively wish to apologize for having misled the generally educated public by spreading ideas about the determinism of systems satisfying Newton's laws of motion that, after 1960, were proved to be incorrect." Sir James was referring to the novelty of low-dimensional deterministic chaos, which had been discovered by E. Lorenz in 1963 and then invented by J. Yorke in 1975. It was Yorke who provided both its name and significance (much as it might be said that in the seventeenth century the novelty of oxygen was discovered by Priestley and then invented by Lavoisier. See T. Kuhn, 1962/1970, Chapter VII).

Note that although the first evidence of nonlinear behaviour – the entrainment of frequency in two coupled oscillators (clocks) – was reported by Huyghens near the middle of the seventeenth century, the productive study of nonlinear systems has been delayed until near the middle of the twentieth. One reason why such studies have languished is that nonlinear dynamics is based on nonlinear differential equations and most differential equations cannot be solved in closed form. For an arbitrary differential equation the so-called analytic solutions cannot be obtained because the repertoire of standard functions (polynomials, trigonometric, exponential, etc.) in terms of which solutions may be expressed is too limited. Moreover, there is a clear fundamental difference between linear and nonlinear equations. Any two solutions of a linear equation can be added together to form a new solution; this is the <u>superposition principle</u>. The powerful analytic methods of Fourier and Laplace transforms depend upon being able to superpose solutions. But for nonlinear equations superposition fails. No general analytic methods exists for solving the general nonlinear equations. The alternative iterative qualitative geometrical methods for the study of nonlinear equations that were introduced by Poincaré in the late 19th century have achieved no small success but really only became generally useful with the advent of computers in the third quarter of the 20th century. Thus, although the origins of nonlinear dynamics, with Poincaré in the late 19th century, antedate those of quantum mechanics, with Heisenberg and Schrödinger in the early 20th century, the development of nonlinear dynamics had to wait on the development of quantum mechanics through which solid state physics could advance far enough to create the "chip" that makes computers possible. It is perhaps of interest that a similar technological constraint has guided the development of statistical methods in this century. The mechanical calculators of the early twentieth century could only "do sums" and sums of squares. This made possible the development of Analysis of Variance methods but retarded that of Regression methods, which are now probably the most widely used statistical methods in science and engineering. The development of regression methods also had to wait on quantum mechanics and its computer "chips". There was little serious astronomy before the telescope, little serious bacteriology before the microscope, and there could be no serious "chaology" before the computer chip.

There is surely fascination (and undeniable frisson) in the ideas of deterministic chaos. It has rightly been referred to as the third scientific revolution in the twentieth century physics, after the quantum and relativity theories (67). But clearly, one must not become obsessed by chaos; the whole of the dynamics of nonlinear systems is of equal fascination and importance, not just the occurrence of chaos, which is only one manifestation of nonlinearity.

There are now three broad frontiers in modern physics: the very small (e.g., elementary particles), the very large (e.g., cosmology), and the very complex (e.g., nonlinear dynamical systems) (68). The phenomena of the very small and the very large are hard to see; those of the very complex are hard to miss: They are all around us. Yet they consistently have been missed until the past thirty years. Or perhaps ignored. (There are several instances of this in the history of chaos.) For, as T. Kuhn (1962/1970) has observed, "Normal science does not aim at novelties of fact or theory and, when successful, finds none ... Normal science, for example, often suppresses fundamental novelties because they are necessarily subversive of its commitments." But one reason for the omission is surely that, "One sees what one knows" (J. Goethe). Or, "When examining normal science we shall want formally to describe that research as a strenuous and devoted attempt to force nature into the conceptual boxes provided by professional education" (T. Kuhn, 1962/1970). Until quite recently "normal science" was "linear science", i.e., the study of shallow waves, low amplitude oscillations, small temperature gradients (69), etc., a practice that may lead to solving the wrong equations in many real-world problems since in nonlinear systems, "More is different" (70). For the real world is, as Ian Stewart has observed, "relentlessly nonlinear," or, as the mathematician Stanislaw Ulam famously remarked, "To speak of nonlinear science is like speaking of non-elephant zoology." But unfortunately, the set of "conceptual boxes" issued to most of us in our professional educations did not include one marked "nonlinear phenomena and models thereof." The aim of the present entry-level Workshop is to begin to address some of these earlier omissions and thereby improve our vision.

PORTRAITS

"Every object or physical form, can be represented as an attractor C of a dynamical system on a space M of internal variables."
...
"All creation and destruction of forms or morphogenesis, can be described by the disappearance of the attractors representing the initial forms, and their replacement or capture by the attractors representing the final forms."
R. Thom, 1972

Some portraits of the strange and violent world of nonlinear dynamical systems follow. Most of the subjects of the portraits are the forced Duffing and Van der Pol oscillators in various settings determined by the system parameters (71). These are prototypical dissipative nonlinear oscillators. The respective equations of motion are

1) Duffing (passive) oscillator: $\ddot{x} + \alpha\dot{x} + \omega_0^2 x + \beta x^3 = \rho \sin\Omega t$.

2) Van der Pol (self-excited) oscillator: $\ddot{x} - \gamma(1 - x^2)\dot{x} + \omega_0^2 x = \rho \sin\Omega t$.

In 1) the nonlinearity resides in the restoring force ($\omega_0^2 x + \beta x^3$); in 2) it resides in the damping force ($\gamma(1-x^2)\dot{x}$). In the above notation, x is the state variable of the system, $\dot{x} = dx/dt$, $\ddot{x} = d\dot{x}/dt$. The control parameters are identified by Greek letters. In the bifurcation diagrams the parameter is the <u>ordinate</u>, the state variable is the <u>abscissa</u>.

The recurrent behaviour of <u>dissipative</u> systems can be characterized by the phase space attractors. These attractors can, in turn, be partially characterized by the Lyapunov exponents. The largest Lyapunov exponent, λ_1, identifies the type of recurrent behaviour: periodic, $\lambda_1 < 0$; quasiperiodic, $\lambda_1 = 0$; aperiodic (chaotic), $\lambda_1 > 0$. On a strange attractor ($\lambda_1 > 0$) two trajectories with initial separation $\delta(0)$ diverge exponentially, $\delta(t) = \delta(0)\exp(\lambda_1 t)$. If the Lyapunov exponents are $\lambda_1 \geq \lambda_2 \geq ... \geq \lambda_m$, then the Lyapunov dimension of the attractor as given by Kaplan-Yorke (39) is

$$D_\lambda = p + \frac{\lambda_1 + \lambda_2 + \ldots \lambda_p}{|\lambda_{p+1}|}$$

where p is the maximum value of i for which $\lambda_1 + \ldots + \lambda_i > 0$. For example, for Fig. 2a

$$D_\lambda = 1 + \frac{0.185}{0.335} = 1.552.$$

In Fig. 2a the <u>limit cycle</u> has Lyapunov exponents $(\lambda_1, \lambda_2) = (-9.975, -0.075)$ and $D_\lambda = 0$; the <u>strange attractor</u> has Lyapunov exponents $(\lambda_1, \lambda_2) = (0.185, -0.335)$ and $D_\lambda = 1.552$. In Fig. 3d the <u>strange attractor</u> has Lyapunov exponents $(\lambda_1, \lambda_2) = (0.062, -0.312)$ and $D_\lambda = 1.198$; the <u>limit cycle</u> has Lyapunov exponents $(\lambda_1, \lambda_2) = (-0.071, -0.179)$ and $D_\lambda = 0$. In Fig. 4a the Lyapunov exponents for the strange attractor are $(\lambda_1, \lambda_2) = (0.094, -5,102)$ and $D_\lambda = 1.018$. In Fig. 6b the Lyapunov exponents for the 2-torus are $(\lambda_1, \lambda_2) = (0, -0.196)$; the attractor dimension is $D_\lambda = 0$. In Fig. 6c the Lyapunov exponents for the period-3 limit cycle are $(\lambda_1, \lambda_2) = (-0.535, -3.126)$; the attractor dimension is $D_\lambda = 0$. In Fig. 6d the Lyapunov exponents for the strange attractor are $(\lambda_1, \lambda_2) = (0.051, -2.776)$; the attractor dimension is 1.019.

Figure 1) Attractors, basins, and bifurcations for a bistable system. The occurrence of a dynamic <u>cusp catastrophe</u> in a Duffing oscillator. Annihilation of limit cycles in saddle-node bifurcations as the parameter Ω is changed gives rise to the hysteresis and "jump" phenomena that characterize a nonlinear resonance (72, 73).

Figure 2) A <u>blue-sky catastrophe</u> in the Duffing oscillator. The death of a <u>strange attractor</u> and its <u>basin</u> as the parameter Ω is changed (74). The strange attractor disappears "into the blue sky" and the system "jumps" to the coexisting limit cycle. The "ghost" of the strange attractor remains.

Figure 3) <u>Explosive bifurcations</u> in passive and self-excited (auto-catalytic) oscillators as the parameter ρ is changed. The birth of a <u>strange attractor</u> from a <u>limit cycle</u> in the Duffing oscillator. The birth of a <u>2-torus</u> from a "mode-locked" (entrained) state in the Van der Pol oscillator. Both transitions occur via saddle-node bifurcations. Like the "jump" phenomenon in the Duffing oscillator, "mode-locking" is unique to nonlinear systems. Neither phenomenon can occur in the absence of nonlinearity (75). Note also that every nonlinear system responds differently to a given form of time-dependent force, and even when the response has been found in any given system it will not simply scale up or down, with the <u>nature</u> of the response unchanged, to an amplification of the driving force. As shown quite dramatically in these figures, the behaviour of a strongly driven nonlinear system may be entirely different from its behaviour under weak driving; a "dose-dependence" feature that must be kept in mind when considering either the design or the deployment of nonlinear systems.

Figure 4) <u>Subtle bifurcations</u>. The death of a strange attractor in a <u>subduction</u>, its rebirth via a cascade of <u>super-critical flip (periodic-doubling) bifurcations</u> and its subsequent abrupt expansion via an <u>interior crisis</u> (collision of the strange attractor with an unstable limit cycle) as the parameter Ω is changed (74, 76, 77). The values of Ω at which the successive flip, or period-doubling, bifurcations occur obey the Feigenbaum scaling rule:

$$\frac{\Omega_i - \Omega_{i-1}}{\Omega_{i+1} - \Omega_i} = \delta$$

where $\delta = 4.6692$ ... Hence, we find from this diagram, the estimate $\hat{\delta}$ where

$$\hat{\delta} = \frac{2.8505 - 2.8459}{2.8515 - 2.8505} = 4.60. \text{ Also, } \hat{\delta}/\delta = 0.9852.$$

Figure 5) <u>Metamorphoses</u> of basin boundaries. The change from smooth to fractal basin boundaries for a Duffing oscillator as the parameter ρ is changed. Two or more basins of attraction are called <u>strange</u> if some region of their boundary is <u>fractal</u>. Cause and effect of "Final state sensitivity" are shown (78, 79, 80).

Figure 6) <u>Subtle bifurcation. Supercritical Niemark (secondary Hopf) bifurcation</u> from a limit cycle to a 2-torus in a van der Pol oscillator. A strange attractor is born via Type I intermittency in a saddle-node bifurcation from the mode-locked period-3 limit cycle created from the torus in a saddle-node bifurcation as the parameter Ω is changed (74, 81, 82).

Figure 7) <u>The geometry of the strange attractors</u> for the Duffing and van der Pol oscillators. Successive enlargements of sections of the attractors showing the characteristic features of these attractors: 1) They are <u>fractals</u>; 2) They are <u>folded</u>. Both features are quite evident for the Duffing oscillator; only the folds are apparent in the strange attractor of the van der Pol oscillator.

Figure 8) <u>Heuristic models of the interaction of a host immune system with a target population (bacteria, virus, tumor cells)</u>. The interaction is modelled in terms of the attractors and basins of a nonlinear dynamical system described by two coupled ordinary differential equations:

$$\dot{x} = rx - kxy. \qquad \dot{y} = p\frac{x^u}{m^v + x^v} + s\frac{y^n}{c^n + y^n} - dy$$

where x is a measure of the target population and y is a measure of the

immune competence of the host. The nonlinearity resides in the last term of the RHS of the first equation and in the first and second terms of the RHS of the second equation. The nonlinear model captures much of the rich dynamical behaviour to be found in the target-host interaction. In phase plots a-d the target population x is plotted along the abscissa and host immune competence y along the ordinate. Several different time courses of disease are described in terms of trajectories in the attractors and basins of the phase space of the dynamical system corresponding to different sets of values of the parameters: r, k, p, u, v, s, n, d, m, c (83, 84).
Figure 9) <u>Entrainment</u> (mode-locking) in a forced van der Pol oscillator. When two dissipative oscillators are coupled together, their combined asymptotic motion can be described by a 2-torus. When the oscillators modelock the combined motion is periodic; the attractor is a limit cycle embedded on the torus. When they are unlocked the combined motion is quasiperiodic; the trajectory is said to be ergodic and densely covers the torus. In the plane of the control parameters (Ω, ρ) for a forced oscillator, the regions of mode-locking are V-shaped bands called Arnold tongues, or horns, each bounded by two arcs of saddle-node bifurcations. The Arnold tongues are immersed in a "sea" of quasiperiodic motion. The location of each tongue is determined by the ratio Ω/ω_0 = p/q where p and q are prime integers. The width of the tongue at a given value of p/q is determined by ρ. For large values of ρ the Arnold tongues overlap implying the co-existence of different periodic oscillations and therefore, chaos. The Arnold tongues are labelled by their winding numbers p/q, which are organized into the <u>self-similar</u> (fractal) Farey sequence. Between Arnold tongues p_1/q_1 and p_2/q_2 is the Arnold tongue $(p_1 + p_2)/(q_1 + q_2)$; i.e., between the tongues at 1/3 and 1/5 lies the tongue at 1/4. Figure 9 is for a weakly nonlinear oscillator, whereas Figure 4 is for a strongly nonlinear oscillator. Note that <u>entrainment</u> is of paramount importance to biological oscillators because it provides a mechanism by which they are coupled to the environment (31).
Figure 10) Symmetry-breaking instability in a dissipative system. The Hopf bifurcation in a chemical oscillator (85, 86, 87). In addition to the van der Pol oscillator, there is another prototypical autocatalytic oscillator, the so-called Brusselator (after its place of origin, The Free University of Brussels). The hypothetical reaction sequence is
1) A $\longrightarrow$ x; 2) B+x $\longrightarrow$ y+D; 3) 2x+y $\longrightarrow$ 3x; 4) x $\longrightarrow$ E. The autocatalytic step, the source of the instability, is at the trimolecular reaction in 3). The onset of symmetry-breaking or a limit cycle always requires an instability. Note that chemical systems are nonlinear when the reaction rates are proportional to products and powers of concentrations of reactants. The Brusselator has a cubic nonlinearity: x^2y. Assuming that the concentrations

of A, B, D, and E are controlled from the outside, i.e., the system is coupled to its environment (a non-equilibrium system), the rate equations for the reactants x and y can be written as the sets of coupled equations

$$dx/dt = A - (Bx - x^2y) - x$$
$$dy/dt = Bx - x^2y.$$

The system has two fixed points at $(x, y) = (A, B/A)$ and a Hopf bifurcation point at $B_c = 1+A^2$. For $B < B_c$ the fixed point is a stable focus; for $B > B_c$, it is an <u>unstable</u> focus and there is a stable limit cycle of amplitude $(B - B_c)^{1/2}$; the frequency is independent of $|B - B_c|$ (to a second approximation). The Brusselator has <u>autocatalytic dynamics</u> if $B > 1 + A^2$, whereas it is dissipative (in the usual sense) if $B < 1 + A^2$. (<u>N.B.</u>: "dissipative" refers to the exchange of energy/matter with the environment of the system. The Hopf bifurcation of the stable fixed point may be written

$$\text{stable fixed point} \begin{cases} \nearrow \text{stable limit cycle} \\ \searrow \text{unstable fixed point} \end{cases}$$

The Hopf bifurcation is an exchange of stability between an attracting fixed point and a limit cycle. (<u>N.B.</u>: A stable limit cycle, as well as a stable fixed point, may undergo a Hopf bifurcation. The result is a 2-torus, as in Figure 6a. Such a transition is more commonly called a Niemark bifurcation.) For $B < B_c$, small perturbations decay as $\exp\{(B - B_c)t\}$; for $B = B_c$, they decay as $1/\sqrt{t}$. (This transition from exponential to algebraic (power-law) decay at a transition point is rather common in many branches of physics; it is known as <u>critical slowing down</u>.) For the Hopf bifurcation the critical point $B = B_c$ is known as a <u>vague attractor</u> (88). There are two types of Hopf bifurcation: subcritical and supercritical. In the former, the onset of oscillation occurs at a finite amplitude, i.e., it is discontinuous, and there is a finite range of the control parameter for which the stable limit cycle and the stable focus co-exist, i.e., the system enters into a bistable stable; the transition exhibits hysteretic discontinuities. For the supercritical Hopf bifurcation, the onset of oscillation occurs at an infinitesimal amplitude, i.e., it is continuous; the system does not enter bistable state and there is no hysteresis.

Note that limit cycles can only occur in nonlinear systems. Moreover, in a chemical chain reaction involving two intermediate reactants the order of at least one reaction step must be trimolecular or higher (86).

"The great conceptual importance of the limit cycle from the thermodynamic point of view comes from its 'ergodicity'. Whatever the initial conditions, the final state is always described by the same periodic trajectory. In this respect, there is a close analogy with the ergodic processes considered in statistical mechanics when the system tends to an equilibrium state, irrespective of the initial condition" (87).

For $B > B_c$ and $A = a_0 + a_1 \cos\Omega t$, the Brusselator will exhibit quasiperiodic or chaotic behaviours depending on the values of a_1 and Ω (89).

The mitotic "clock" can be modelled as a continuous biochemical oscillator that "gates" mitosis and DNA replication. However, these latter processes are not themselves causal parts of the oscillator. An early model of the "clock" was based on a modification of the positive feedback (autocatalytic) Brusselator (90). More recent results with a negative feedback oscillator "... support the inclusion ... of the mitotic control system among the chemical and biological processes capable of nonequilibrium self-organization in the form of limit cycle oscillatory behaviour" (91).

If the Brusselator model is augmented by terms representing diffusion of the intermediate species x and y (by adding $D_x d^2x/dr^2$ and $D_y d^2y/dr^2$ to the RHS of equations for dx/dt and dy/dt, respectively, where D_x, D_y are diffusion coefficients), the augmented model describes a reaction-diffusion dynamical system that gives rise to spatial as well as temporal patterns (13). The spatial pattern, a dissipative structure, is maintained by competition between reaction and diffusion processes. In analogy with the Hopf bifurcation that gave rise to the limit cycle, Prigogine (13) has called the spatial symmetry-breaking bifurcation a Turing bifurcation.

Note that both the limit cycle and the Hopf bifurcation (more accurately, the Poincaré-Andronov-Hopf bifurcation) were both discovered /invented by Poincaré in his work on the three-body problem at the end of the 19th century.

The foregoing portraits illuminate, in Winfree's phrase, "a study of temporal morphology, of shapes not in space so much as in time" (92). The portraits provide vivid illustrations of two important and characteristic features of driven nonlinear oscillators: 1) The response of the oscillator exhibits a rich spectrum of recurrent behaviours: equilibrium, periodic, quasiperiodic, or chaotic. 2) Which of these behaviours are observed for a given oscillator depends sensitively on the initial conditions and on the amplitude ρ and frequency Ω of the driver. Unlike linear oscillators, only a small change in the initial conditions or in ρ or Ω may produce very large changes — both quantitative and qualitative — in the response. The changes in behaviour for changes in the parameters or initial conditions may be quite abrupt, in fact, even discontinuous, or explosive.

There are two interesting - and not unrelated - consequences of the presence of nonlinearity in physical systems:
1) "Safety" is (usually) an empirical science in, e.g., a) Bridges, b) Aircraft,

c) Reactors.

2) In "extrapolation" of mechanistic models, of either physical or biological systems, it is (usually) the case that the "mechanism" will change for observations outside of the construction set, hence the cautions against any extrapolation that appear in all texts on modelling.

The portraits were made with the following DOS software packages: 1) <u>DYNAMICS. Numerical Explorations</u>. H.E. Nusse and J.A. Yorke, Springer-Verlag, New York, 1994. 2) <u>PHASER. Differential and Difference Equations through Computer Experiments</u>. 2nd ed., H. Koçak, Springer-Verlag, New York, 1989.

The American Institute of Physics (AIP) also offers DOS software for nonlinear dynamics: 1) <u>CHAOS DEMONSTRATIONS</u>. IBM PC Version 3.0, 1992. 2) <u>CHAOS DATA ANALYZER</u>. IBM PC Version 1.0, 1992. Both can be purchased from the Academic Software Library, Box 8202, North Carolina State University, Raleigh, NC 27695-8202. The first package includes excellent demonstrations of fractal structures (e.g., diffusion-limited aggregations) and fractal processes (e.g., "1/f" noise).

The early observations of the "very small" and the "very large" were made possible by the silica lens. The early observations of the "very complex" are made possible by the silicon chip.

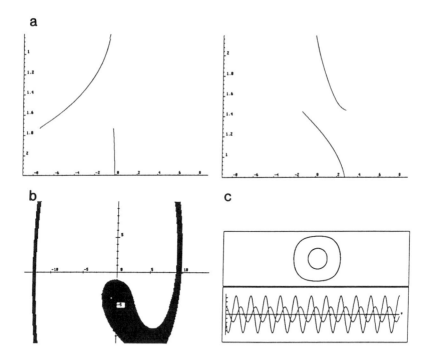

FIGURE 1a. Bifurcation diagram (x, Ω). Saddle-node bifurcations between co-existing periodic states (limit cycle attractors) with different amplitudes and phases occur at $\Omega = \Omega_u$ for Ω increasing (left panel) and at $\Omega = \Omega_l < \Omega_u$ for Ω decreasing (right panel); a hysteresis phenomenon. At each bifurcation a stable and an unstable limit cycle collide and are annihilated; the system "jumps" to the other co-existing stable limit cycle. These are dangerous discontinuous bifurcations.

FIGURE 1b. Basin diagram (x, $\dot{x}$) for the oscillator in Fig. 1a with co-existing point attractors. All trajectories with initial conditions in the white region asymptote to one point (limit cycle) attractor. All trajectories with initial conditions in the black region asymptote to the other point (limit cycle) attractor.

FIGURE 1c. The upper panel is the superposition of the phase portraits of the two co-existing limit cycles. The lower panel is the super-position of the respective corresponding time-series.

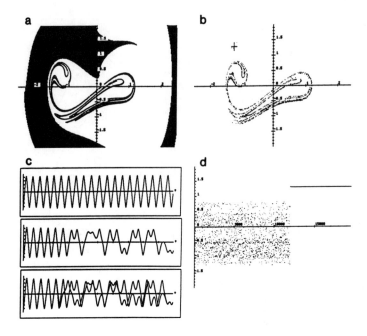

FIGURE 2a. Basin diagram (x, ẋ) of co-existing periodic and aperiodic (chaotic) attractors for the forced Duffing oscillator. The respective basins are black and white. The Poincaré sections of the two attractors, a limit cycle (point) and a strange attractor (fractal object) are superimposed. The strange attractor has fractal dimension D_λ = 1.552.

FIGURE 2b. A Poincaré section of the limit cycle attractor and the "ghost" of the strange attractor following a "blue-sky catastrophe" in which, as the parameter Ω is changed to a value slightly less than that of Fig. 2a, the strange attractor collided with the basin boundary. Both the attractor and its basin were annihilated and disappeared from the phase space of the oscillator; only the limit cycle (cross) of Fig. 2b remains. The presence of the "ghost" gives rise to "transient chaos" as shown in Fig. 2d.

FIGURE 2c. The respective time-series (x,t) of the two co-existing attractors of Fig. 2a. Upper panel − limit cycle. Middle panel − strange attractor. Bottom panel − a superposition of the respective time-series for two initial conditions on the strange attractor differing by less than 0.05% and showing the sensitivity to initial conditions and predictability in the short run but not in the long that is a signature of deterministic chaos.

FIGURE 2d. Time-series for a trajectory with initial conditions in the region of the "ghost" of the strange attractor for the Duffing oscillator with parameters of Fig. 2c. The ordinate is ẋ. An initial chaotic transient, precedes the final periodic state.

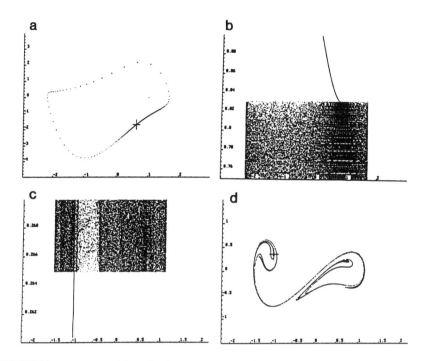

FIGURE 3a. A super-position of Poincaré sections $(x, \dot{x})$ of a forced Van der Pol oscillator for quasi-periodic and periodic (mode-locked) conditions corresponding to different values of the parameter ρ. The respective attractors are the 2-torus and limit cycle. The latter is identified by the cross; it is clearly a subset of the torus.

FIGURE 3b. The bifurcation diagram (x, ρ) for a forced Van der Pol oscillator. The control parameter is the amplitude, ρ, of the forcing. There is a saddle-node bifurcation from the limit cycle (mode-locked state) to the 2-torus (quasi-periodic state) at $\rho \cong 0.83$.

FIGURE 3c. The bifurcation diagram (x, ρ) for a forced Duffing oscillator. The control parameter is the amplitude, ρ, of the forcing. There is a saddle-node bifurcation from the limit cycle to the strange attractor, fractal dimension $D_\lambda = 1.198$ at $\rho \cong 0.265$.

FIGURE 3d. A super-position of Poincaré sections $(x, \dot{x})$ of the forced Duffing oscillator for periodic and aperiodic (chaotic) conditions for two different values of the parameter ρ. The respective attractors are the limit cycle and the strange attractor. The former is identified by the cross. Here, as in Figure 3a, at the threshold value of ρ the attractor jumps discontinuously in size, i.e., these are <u>explosive discontinuous bifurcations</u>.

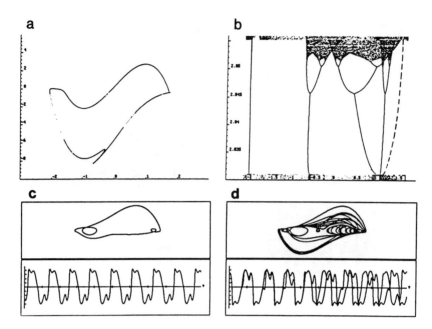

FIGURE 4a. The Poincaré section of a strange attractor for a forced Van der Pol oscillator at $\Omega = 2.856$ (near top of Fig. 4b). The closed curve is <u>not</u> homeomorphic (topologically equivalent) to a circle. The "spike" on the lower arc identified the presence of <u>folding</u> that is characteristic of a strange attractor. The attractor has a fractal dimension $D_\lambda = 1.018$.

FIGURE 4b. Bifurcation diagram in a period-4 window for the Van der Pol oscillator for the frequency region $2.830 \leq \Omega \leq 2.856$. There is a <u>subduction</u> of a strange attractor in <u>saddle-node</u> bifurcations at $\Omega \sim 2.830$ (near bottom of Fig. 4b) giving rise to stable and unstable periodic oscillations of period-4. The unstable orbits are shown as dashed lines. The stable oscillation undergoes a cascade of flip (period doubling) bifurcations to the strange attractor of Fig. 4a as Ω increases. As $\Omega \sim 2.855$ the strange attractor collides with the unstable orbit in an <u>interior crisis</u> in which the size of the strange attractor increases in an <u>explosive bifurcation</u>.

FIGURE 4c. The upper panel is the phase portrait of the **period-4** limit cycle of Fig. 4b. The lower panel gives the corresponding time-series.

FIGURE 4d. The upper panel is the phase portrait of the strange attractor of Fig. 4a. The lower panel gives the superposition of two corresponding time-series with nearly identical initial conditions, showing the presence of deterministic chaos. Note that the strange attractor is symmetric, but the limit cycle of Fig. 4c is asymmetric.

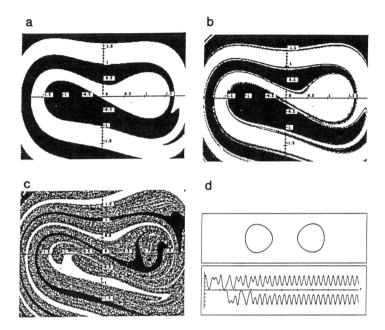

FIGURES 5a-c. Metamorphosis of a basin boundary. The basins of attraction for a forced Duffing oscillator are shown for three different values of the control parameter, ρ. The value of ρ increases from a through c. For each value of ρ there are two co-existing limit cycle attractors. For one attractor the basin is black; for the other the basin is white. For a the basin boundary is smooth; for b and c the boundary is fractal. Figure c shows a Cantor-set type structure.

FIGURE 5d. As a consequence of fractal basin boundaries the prediction of the final dynamical state of the systems is very difficult, i.e., there is "final state sensitivity" (Grebogi, C. et al, 1983). The upper panel shows the respective limit cycles for the two co-existing final states corresponding to the two basins of Figures 5a-c. The lower panel shows the time-series from two different initial conditions in the Cantor-set region differing by less than 0.1%, demonstrating the difficulty in predicting the final state of the oscillator at this set of parameter values. Fractal basin boundaries, as well as strange attractors, commonly occur in nonlinear dynamical systems.

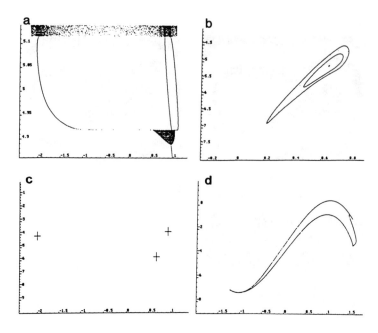

FIGURE 6a. Bifurcation diagram (x, Ω). Supercritical Niemark (secondary Hopf) bifurcation from a period-1 limit cycle to 2-torus in the lower part of the figure. As Ω is increased the two frequencies become commensurate and mode-lock via a saddle-node bifurcation to give the period-3 limit cycle shown in the center of the figure. The period-3 limit cycle is replaced by a strange attractor via a saddle-node bifurcation (Type I intermittency) in the upper part of the figure.

FIGURE 6b. A Superposition of two Poincaré sections (x, ẋ) through the 2-torus (closed curves) and a single section through the period-1 limit cycle (single point) at the bottom of Figure 6a.

FIGURE 6c. Poincaré section (x, ẋ) of the period-3 limit cycle in center of Fig. 6a.

FIGURE 6d. Poincaré section (x, ẋ) of the strange attractor at the top of Fig. 6a. The strange attractor has fractal dimension D_λ = 1.018. The "spike" on the upper arc identifies the presence of <u>folding</u> that is characteristic of a strange attractor.

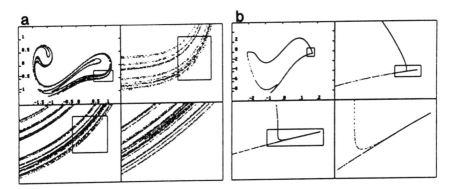

FIGURE 7a. Successive enlargements of a section of the strange attractor of the Duffing oscillator of Fig. 2a showing the <u>fractal</u> structure that is characteristic of chaotic dynamics. The largest Lyapunov exponent is $\lambda_1 = 0.185$. The fractal dimension is $D_\lambda = 1.552$.

FIGURE 7b. Successive enlargements of a section of the strange attractor of the van der Pol oscillator of Fig. 4a showing the <u>folded</u> structure that is characteristic of chaotic dynamics. The largest Lyapunov exponent is $\lambda_1 = 0.094$. The fractal dimension is $D_\lambda = 1.018$. (Note that for the Lorenz strange attractor, the "butterfly," the largest Lyapunov exponent is $\lambda_1 = 0.884$. The fractal dimension is $D_\lambda = 2.061$.)

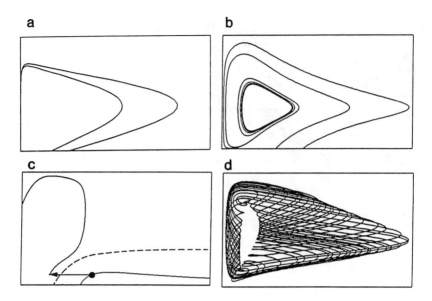

FIGURE 8a. Trajectories in the phase plane describe the successful response of the host to two challenges from target populations of different sizes; both challenges lie within the basin of attraction for cure, both trajectories converge to a point attractor at $x = 0, y > 0$ (targets destroyed). The figure shows why, "it usually gets worse before it gets better."

FIGURE 8b. Limit cycle attractor describing the co-existence of host and target population as in a chronic periodic disease, e.g., herpes simplex, malaria, etc.

FIGURE 8c. Basins of cure and lethality. Therapeutic interventions (e.g., antitumor agents, antibiotics) reduce the target burden and thus move (horizontal arrow) the trajectory across the separatrix (dashed) from the basin of the latter to that of the former.

FIGURE 8d. Attractor for a model of a tumor-host interaction: natural killer (NK) cell number (y-axis) vs. tumor size (x-axis) assuming the tumor production rate r varies sinusoidally. The number of NK cells and the tumor burden fluctuate erratically over time exhibiting an apparently "chaotic" behaviour. Note that the tumor burden fluctuates irregularly even in the absence of treatment.

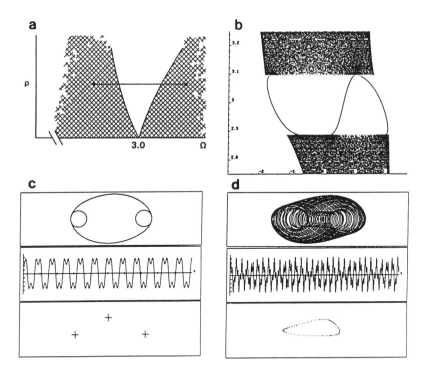

FIGURE 9a. Arnold tongue at $\Omega/\omega_0 = 3.0$ (subharmonic entrainment) in the control space (Ω, ρ) diagram. The horizontal line is at the level of the bifurcation diagram of Fig. 9b. The Arnold tongue is the open wedge-shaped region of periodic (mode-locked) oscillation. Quasi-periodic oscillation occurs in the cross-hatched region.

FIGURE 9b. Bifurcation diagram (x, Ω) through an Arnold tongue for a period-3 limit cycle. The two central lines, say $\Omega = \Omega_l$ and $\Omega = \Omega_u$, are definite edges of the corresponding Arnold tongue, at which there occur, as Ω is varied, saddle-node bifurcations between the periodic and quasi-periodic regions of the oscillator.

FIGURE 9c. Periodic motion in the mode-locked region of Fig. 9a. Upper panel — the phase portrait in the $(x, \dot{x})$ plane. Middle panel — time-series. Lower panel - Poincaré section.

FIGURE 9d. Quasiperiodic motion for Fig. 9a. Upper panel — phase portrait. Middle panel — time-series. Lower panel — Poincaré section.

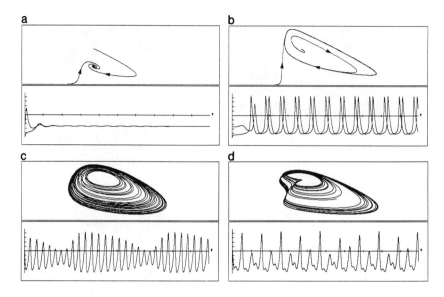

FIGURE 10a. The upper panel presents, in (x, y) space of the Brusselator, a superposition of the fixed point, a stable focus at $(x, y) = (A, B/A)$ and the trajectories for two initial conditions for $B < B_c = 1 + A^2$. The lower panel presents the superposition of the corresponding time-series (x, t).

FIGURE 10b. The upper panel presents, in the (x, y) space of the Brusselator, the superposition of the fixed point, an <u>un</u>stable focus at $(x, y) = (A, B/A)$, a limit cycle and the trajectories for two initial conditions for $B > B_c = 1 + A^2$. There is a supercritical Hopf bifurcation at $B = B_c = 1 + A^2$. The lower panel presents a superposition of the respective time-series (x, t). Note that the amplitude and period of the oscillations for the limit cycle are independent of the initial conditions. Comparison of the respective time-series as $t \longrightarrow \infty$ for the stable focus; (Figure 10a) and limit cycle (Figure 10b) shows the Hopf bifurcation to be a symmetry-breaking bifurcation: the temporal symmetry changes from uniform translation to periodic.

FIGURE 10c. The upper panel presents the phase portrait — a 2-torus in (x, y) space — for a quasiperiodic behaviour of the Brusselator. Changes in either the amplitude a_1 or frequency Ω of the driver will lead to a mode-locked state via a saddle-node bifurcation on the arc of an Arnold tongue. The lower panel presents the corresponding time-series (x, t).

FIGURE 10d. The upper panel presents the phase portrait — a strange attractor in (x, y) space — for a chaotic behaviour of the Brusselator. Note the presence of a fold in the attractor. The lower panel shows the corresponding time-series (x,t).

REFERENCES

1. Spengler, O. *The Decline of the West*. Volume One. A. Knopf. 1932, p 59.
2. Hobsbawm, E. *The Age of Extremes. A History of the World, 1914-1991*. Pantheon. NY. 1994. pp 550-551.
3. Hayles, N.K. Chaos as Orderly Disorder: Shifting Ground in Contemporary Literature and Science. *New Literary History*. 20: 305-322. 1989.
4. Kuhn, T. *The Structure of Scientific Revolutions*. Univ. Chicago Press. 1962/1970.
5. Yates, F.E. *Self Organizing Systems. The Emergence of Order*. Plenum Press. NY. 1987. p 210.
6. Landau, L. and Lifschitz, E. *Statistical Physics*. Addison-Wesley. Reading, MA. 1958. Chapter XIV.
7. Nayfeh, A.H. and Mook, D.T. *Nonlinear Oscillations*. John Wiley. NY. 1979. Chapters 1 and 4.
8. Hayashi, C. *Nonlinear Oscillations in Physical Systems*. Princeton Univ. Press. New Jersey. 1964/1985. Chapter 12. (See Figs. 12.11 & 12.12.)
9. Minorsky, N. *Introduction to Nonlinear Mechanics. Topological Methods – Analytical Methods. Nonlinear Resonance – Relaxation Oscillators*. J.W. Edwards. Ann Arbor. 1947. Chapter XVIII.
10. Thom, R. *Structural Stability and Morphogenesis. An Outline of a General Theory of Models*. Addison-Wesley. Reading, MA. 1972/1989. Chapter 1.
11. Haken. H. *Synergetics. An Introduction. Nonequilibrium Phase Transition and Self-Organization in Physics, Chemistry, and Biology*. 2nd ed. Springer-Verlag. NY. 1978. Chapter 1.
12. Nicolis, G. and Prigogine, I. *Self-Organization in Nonequilibrium Systems. From Dissipative Structures to Order Through Fluctuations*. John Wiley, NY. 1977. Chapters 5-7. (See also Nicolis, G. and Prigogine, I. *Exploring Complexity*. W.H. Freeman. NY.1989.)
13. Prigogine, I. Time, Structure, and Fluctuations. *Science*. 201: 777-785. 1978.
14. Jackson, E.A. *Perspectives in Nonlinear Dynamics. Vol. 1*. Cambridge Univ. Press. NY. 1991. Chapter 5.
15. Anderson, P.W. and Stein, D.L. Broken Symmetry. Emergent Properties, Dissipative Structures, Life. Are They Related? *Self-Organizing Systems. The Emergence of Order*. F.E. Yates, ed. Plenum Press. NY. 1987.
16. Gilmore, R. *Catastrophe Theory for Scientists and Engineers*. John Wiley. NY. 1981. Chapter 6.
17. Cobb, L. Statistical Catastrophe Theory. IN: *Encyclopedia of Statistical Sciences*. 8: 634-640. 1988.
18. Cobb, L. and Zacks, S. Applications of Catastrophe Theory for Statistical Modelling in the Biosciences. *J. Am. Stat. Assoc.* 80: 793.802. 1985.
19. Poston, T. and Stewart, I. *Catastrophe Theory and Its Applications*. Pitman Pub. London. 1978. p 419. (See also Attneave, F. Multistability in Perception. *Scientific American*. 225: 62-71. 1971.)
20. Thom, R. *Structural Stability and Morphogenesis. An Outline of a General Theory of Models*. Addison-Wesley. Reading, MA. 1972/1989.
21. Stewart, I. and Golubitsky, M. *Fearful Symmetry. Is God a Geometer?* Blackwell. Cambridge, MA. 1992. p 245.

22. Mandelbrot, B.B. *The Fractal Geometry of Nature.* W.H. Freeman. NY. 1983. Chapter I.
23. Kadanoff, L.P. Fractals: Where's the Physics? *Phys. Today.* 39: 6-8. 1986.
24. Strange, G. Wavelet Transforms Versus Fourier Transforms. *Bull. Am. Math. Soc.* 28(2): 288-305. 1993.
25. Falconer, K. *Fractal Geometry Mathematical Foundations and Applications.* John Wiley. NY. 1990. Chapter 3 and 9.
26. Bassingthwaighte, J., Liebovitch, L. and West, B. *Fractal Physiology.* Oxford Press. NY. 1994. Chapter 2.
27. Feder, J. *Fractals.* Plenum Press. NY. 1988. Chapter 3.
28. Sørensen, E.S., Fogedby, H.D. and Mouritsen, O.G. Crossover from Nonequilibrium Fractal Growth to Equilibrium Compact Growth. *Phys. Rev. Lett.* 61: 2770-2773. 1988.
29. Thom. R. *Structural Stability and Morphogenesis. An Outline of a General Theory of Models.* Addison-Wesley. Reading, MA. 1972/1989. Chapter 4.
30. Luo, X. and Schramm, D. Fractals and Cosmological Large Scale Structure. *Science.* 256: 513-514. 1992.
31. Schroeder, M. *Fractals, Chaos, Power Laws.* Freeman. NY. 1991. Chapter 5.
32. Voss, R.F. Random Fractals: Self-Affinity in Noise, Music, Mountains, and Clouds. *Physica D.* 38: 362-371. 1989.
33. Bak, P., Tang, C. and Wiesenfeld, K. Self-organized Criticality. *Phys. Rev. A.* 38: 364-374. 1988.
34. Turcotte, D. *Fractals and Chaos in Geology and Geophysics.* Cambridge Univ. Press. NY. 1992. Chapter 16.
35. Lorenz, E.N. Deterministic Nonperiodic Flow. *J. Atmos. Sci.* 20: 130-141. 1963.
36. Pippard, A.B. *Response and Stability.* Cambridge Univ. Press. NY. 1985.
37. Thompson, J.M.T. and Stewart, H.B. *Nonlinear Dynamics and Chaos.* John Wiley. NY. 1986. Chapter 7.
38. Strogatz, S. *Nonlinear Dynamics and Chaos.* Addison-Wesley. NY. 1994. Chapter 3.
39. Nayfeh, A.H. and Balachandran, B. *Applied Nonlinear Dynamics. Analytical, Computational, and Experimental Methods.* John Wiley. 995. Chapter 5.
40. Grebogi, C., Ott, E. and Yorke, J.A. Metamorphoses of Basin Boundaries in Nonlinear Dynamical Systems. *Phys. Rev. Lett.* 56: 1011-1014. 1986.
41. Grebogi, C., McDonald, S., Ott, E. and Yorke, A. Final State Sensitivity: An Obstruction to Predictability. *Phys. Lett.* 99A: 415-418. 1983.
42. Nicolis, G. and Prigogine, I. Symmetry Breaking and Pattern Selection in Far-from Equilibrium Systems. *Proc. Natl. Acad. Soc. USA.* 78: No. 2. 659-663. 1981.
43. Thompson, J.M.T. and Stewart, H.B. *Nonlinear Dynamics and Chaos.* John Wiley. NY. 1986. p 348.
44. Sugihara, G. and May, R.M. Nonlinear Forecasting as a Way of Distinguishing Chaos from Measurement Error in Time Series. *Nature.* 344: 734-741. 1990.
45. Ott, E., Grebogi, C. and Yorke, J.A. Controlling Chaos. *Phys. Rev. Lett.* 64: 1196-1199. 1990.
46. Ditto, W.L., Rauseo, S.N. and Spano, M.L. Experimental Control of Chaos. *Phys. Rev. Lett.* 65: 3211-3214. 1990.
47. Gorman, M. and Robbins, K.A. Real-time Identification of Flame Dynamics. IN: *Applied Chaos.* J.H. Kim and J. Stringer, eds. John Wiley. NY. 1992. pp 261-276.

48. Vallis, G. El Niño: A Chaotic Dynamical System. *Science.* 232: 243-245. 1986.
49. Tziperman, E., Stone, L., Cane, M. and Jarosh, H. El Niño Chaos: Overlapping of Resonances Between the Seasonal Cycle and the Pacific Ocean Oscillator. *Science.* 264: 72-74. 1994.
50. Jin, F., Neelin, J.D. and Ghil, M. El Niño on the Devil's Staircase: Annual Subharmonic Steps to Chaos. *Science.* 264: 70-72. 1994.
51. Bassingthwaighte, J.B., Liebovitch, L.S. and West, B.J. *Fractal Physiology.* Oxford Univ. Press. 1994. Chapter 13.
52. Henon, M. and Heiles, C. The Applicability of the Third Integral of Motion: Some Numerical Experiments. *Astron. J.* 73-79. 1964.
53. Wisdom, J. Urey Prize Lecture: Chaotic Dynamics in the Solar System. *Icarus.* 72: 241-275. 1987.
54. Wisdom, J. Is the Solar System Stable? and Can we Use Chaos to Make Measurements? IN: *Chaos. Soviet-American Perspectives on Nonlinear Science.* D. Campbell, ed. Amer. Inst. Phys. NY. 1990. pp 275-303.
55. Laskar, J. The Chaotic Motion of the Solar System. A Numerical Estimate of the Size of the Chaotic Zones. *Icarus.* 88: 266-291. 1990.
56. Gaponov-Grekhov, A.V. and Babinovich, M.I. *Nonlinearities in Action. Oscillations, Chaos, Order, Fractals.* Springer-Verlag. NY. 1992. Chapter 2.
57. Whitham, G. *Linear and Nonlinear Waves.* John Wiley, NY. 1974. Chapter 1.
58. Brigg, J. and Peat, F.D. *Turbulent Mirror. An Illustrated Guide to Chaos Theory and the Science of Wholeness.* Harper & Row. NY. 1990. p 123.
59. Dodd, R., Elibeck, J., Gibbon, J. and Morris, H. *Solitons and Nonlinear Wave Equations.* Academic Press. NY. 1982. Chapter 1.
60. Zabusky, N. and Kruskal, M. Interaction of "Solitons" in a Collisionless Plasma and The Recurrence of Initial States. *Phys. Rev. Lett.* 15(6): 240-243. 1965.
61. Fermi, E., Pasta, J. and Ulam, S. Studies of Nonlinear Problems. Technical Report LA-1940. Los Alamos Sci. Lab. 1955.
62. Hirato, R. and Suzuki, K. Studies in Lattice Solitons by Using Electrical Networks. *J. Phys. Soc. Japan.* 28: 1366-1367. 1970.
63. Campbell, D. Introduction to Nonlinear Phenomena. IN: *Lectures in the Science of Complexity.* D. Stein, ed. Addison-Wesley. Reading, MA. 1989. pp 63-84.
64. M. Nezlin. Some Remarks on Coherent Structures out of Chaos in Planetary Atmospheres and Oceans. *Chaos.* 4(2): 100-111. 1994.
65. Gigerenzer, G., et al *The Empire of Chance. How Probability Changed Science and Everyday Life.* Cambridge Univ. Press. 1989. p 164.
66. Lighthill, J. The Recently Recognized Failure of Predictability in Newtonian Dynamics. *Proc. Roy. Soc. London. A.* 407: 35-50. 1986.
67. Ford, J. How Random is a Coin Toss?" *Phys. Today.* 40-47. 1983.
68. Davies, P. *The New Physics.* Cambridge Univ. Press. 1989. p 4.
69. Stewart, I. *Does God Play Dice? The Mathematics of Chaos.* Blackwell. NY. 1989. p 83.
70. Anderson, P. More is Different. *Science.* 177: 393-396. 1972.
71. Hale, J.K. and Koçak, H. *Dynamics and Bifurcations.* Springer-Verlag. NY. 1991. pp 497-504.
72. Pippard, A. *Response and Stability. An Introduction to the Physical Theory.* Cambridge Univ. Press. NY. 1985. Chapter 4.
73. Thompson, J. and Stewart, H. *Nonlinear Dynamics and Chaos. Geometrical Methods*

74. *for Scientists and Engineers.* John Wiley. NY. 1986. Chapter 7 (See Fig. 7.9).
74. Nayfeh, A. and Balachandran, B. *Applied Nonlinear Dynamics. Analytical, Computational, and Experimental Methods.* John Wiley. NY. 1995. Chapter 5.
75. Thompson, J. and Stewart, H. *Nonlinear Dynamics and Chaos. Geometrical Methods for Engineers and Scientist.* John Wiley, NY. 1986. Chapter 13 (See Figs. 13.2 & 13.7).
76. Grebogi, C., Ott, E. and Yorke, J. Crises, Sudden Changes in Chaotic Attractors, and Transient Chaos. *Physics D.* 7: 181-200. 1983.
77. Parlitz, U. and Lauterborn, W. Period-doubling Cascades and Devil's Staircase of the Driven Van der Pol Oscillator. *Physical Review A.* 36(3): 1428-1434. 1987.
78. Pezeshki, C. and Dowell, E. On Chaos and Fractal Behaviour in a Generalized Duffing's System. *Physica D.* 32: 194-209. 1988.
79. Grebogi, C., McDonald, S., Ott, E. and Yorke, J. Final State Sensitivity: An Obstruction to Predictability. *Physics Letters.* 99A: 415-418. 1983.
89. Grebogi, C., Ott, E. and Yorke, J.A. Metamorphoses of Basin Boundaries in Nonlinear Dynamical Systems. *Phys. Rev. Lett.* 56: 1011-1014. 1986.
81. Pomeau, Y. and Manneville, P. Intermittent Transition to Turbulence in Dissipative Dynamical Systems. *Comm. Math. Phys.* 74: 189-197. 1980.
82. Thompson, J.M.T. and Stewart, H.B. *Nonlinear Dynamics and Chaos. Geometrical Methods for Engineers and Scientists.* John Wiley. NY. 1986. Chapter 8.
83. Mayer, H., Zaenker, K. and Van der Heiden. A Basic Mathematical Model of the Immune Response. *Chaos.* 5(1): 155-161. 1995.
84. Kuznetsov, V.A., Makalkin, I.A., Taylor, M.A. and Perelson, A.S. Nonlinear Dynamics of Immunogenic Tumors: Parameter Analysis and Global Bifurcation Analysis. *Bull. Math. Biol.* 56(2): 295-321. 1994.
85. Prigogine, I. and Lefever, R. Symmetry Breaking Instabilities in Dissipative Systems. II. *Journal of Chem. Phys.* 48: 1695-1700. 1968.
86. Nicolis, G. and Prigogine, I. *Self-Organization in Nonequilibrium Systems.* John Wiley & Sons. NY. 1977.
87. Glansdorff, P. and Prigogine, I. *Thermodynamic Theory of Structure, Stability and Fluctuations.* John Wiley & Sons. NY. 1971.
88. Mees, A.I. *Dynamics of Feedback Systems.* John Wiley & Sons. NY. 1981.
89. Tomita, K. Chaotic Response of Nonlinear Oscillators. *Physics Reports.* 86: 113-167. 1982.
90. Tyson, J. and Kaufman, S. Control of Mitosis by a Continuous Biochemical Oscillation. *J. Math. Biol.* 1: 289-310. 1975.
91. Goldbeter, A. A Minimal Cascade Model for the Mitotic Oscillator Involving Cyclin and cdc2 Kinase. *Proc. Natl. Acad. Sci. USA.* 88: 9107-9111. 1991.
92. Winfree, A.T. *The Geometry of Biological Time.* Springer-Verlag. 1990. p1.

Catastrophe Theory:

What It Is - Why It Exists - How It Works

Robert Gilmore

Department of Physics & Atmospheric Science
Drexel University, Philadelphia, PA 19104

Abstract. Push on something. It will move. Push a little bit harder and it will move a little bit more. But occasionally a little extra push will produce an extra large response. Such extra responses are called "catastrophes." This kind of behavior is summarized by the phrase "... the straw that broke the camel's back."

Situations in which a gradually increasing force leads to a gradually increasing response, followed by a sudden catastrophic jump to a qualitatively different state, are all too common. They are seen, for example, in the collapse of a bridge, dam, or building; the loss of stability of an aircraft or ship; phase transitions in fluids or solids; ignition of a laser; sudden changes in the earth's climate.

In each of the instances above, and very many others besides, it is possible to see a smooth response and a discontinuous response to a smoothly changing force. It might seem that "Every smooth response is the same, each discontinuous response is discontinuous in its own fashion." In fact, every smooth process is conveniently described by a linear response function or tensor. However, it is not true that every discontinuity is different. Discontinuities are described by mathematical functions called catastrophe functions. Only a relatively small number of such functions exist. These have been extensively studied. They can be constructed for any physical system. The simplest of these have been studied extensively. The result is that only a handful of different types of discontinuities are typically encountered, and each of these exhibits a characteristic set of properties.

A parallel realization that a relatively small number of features characterizes the complex behavior exhibited by nonlinear dynamical systems (sets of ordinary differential equations) has catalyzed the rapid growth of interest in such systems in the last two decades.

In the following sections we describe catastrophe theory. In particular, we describe what it is, why it exists, and how it works.

WHAT IT IS

The Program of Catastrophe Theory

There has been a great deal of confusion over what constitutes 'catastrophe theory.' This came about because the term has been applied in two different ways. In a very broad context, catastrophe theory is the study of solutions of equations, and in particular how these solutions change their properties, and even their numbers, as the parameters which occur in these equations change. There is very little that can currently be said about this broad subject.[1,2]

Elementary catastrophe theory is a much more restricted subject. The equations under study in this case are gradient dynamical systems

$$dx_i/dt = -\partial V/\partial x_i, \quad V = V(x;c) \tag{1}$$

where V is a potential depending on n state variables $x \in R^n$ and k control parameters $c \in R^k$. More specifically, elementary catastrophe theory is a study of how the equilibria ($\partial V/\partial x_i = 0$) of these potentials move about, coalesce and are destroyed, or are created in bifurcations, as the control parameters c are varied. A great deal can be said about elementary catastrophes.

Elementary catastrophe theory occurs at the crossroads of two great mathematical pathways. One is elementary, the other is not. The first path involves a series of results from elementary calculus: the Implicit Function Theorem, the Morse Lemma, and the next stage (the Thom Splitting Theorem). The second path is the search for canonical forms for functions, mappings, processes, ... in the neighborhood of singularities.

Three Theorems of Elementary Calculus

The local properties of a smooth function in any neighborhood are determined by the coefficients of its Taylor series expansion. Depending on the values of these coefficients, one of three theorems is applicable. Two of these theorems are well known; the third is not. All are elementary, and may be taught in introductory calculus courses. The first two theorems are the Implicit Function Theorem and the Morse Lemma. The third is the Thom Splitting Theorem. In the statements below the original coordinate system is x, the new coordinate system is $y = y(x)$, and there is a smooth nonsingular (invertible) transformation in the neighborhood of the point where the Taylor series of the function $f(x)$ is computed.

Implicit Function Theorem. If $f(x)$ ($x \in R^n$) is a smooth function and $Df \neq 0$ at x_o, then there is a smooth change of variables such that

$$f = y_1 \tag{2}$$

in the neighborhood of x_o. Here y_1 is the first coordinate of y.

The Implicit Function Theorem is the basis for all the results relating to physical problems of the type: for a small push there is a small response. It is the rigor behind the statement that "Every smooth process is the same." The Implicit Function Theorem forms the basis for linear response theory, the linearization of complicated but smooth transformations, the existence of linear susceptibility tensors.

But what happens when $Df = 0$ at a point? One might be tempted to conclude that the Implicit Function Theorem is not true. This is not so: theorems are never false - the conditions for which the theorem is true are not satisfied. In this case there is a backup theorem which depends on the coefficients of the second degree terms in the Taylor series expansion of f.

Morse Lemma. If $f(x)$ is a smooth function with $Df = 0$ at x_o and $Det\, D^2 f = Det\, (\partial^2 f / \partial x_i \partial x_j) \neq 0$ at $x = x_o$, then there is a smooth change of variables so that

$$f = -y_1^2 - \cdots - y_j^2 + y_{j+1}^2 + \cdots + y_n^2 = M^n{}_j(y) \tag{3}$$

in the neighborhood of x_o. The quadratic form $M^n{}_j(y)$ is called a Morse j-saddle. It has j unstable directions and $n-j$ stable directions.

The Morse Lemma provides the rigorous basis for a standard approach to the analysis of equilibria: expand the potential around the equilibrium to second order and neglect higher order terms. The constant (0th degree term) is unimportant and can be removed by a shift of origin; the linear terms are absent at an equilibrium, and terms of degree three and higher can be removed by a smooth change of variables. This leaves only quadratic terms, which can be transformed to Morse canonical form by a simple matrix transformation.

What happens when neither the Implicit Function Theorem nor the Morse Lemma is applicable? In this case there is a theorem due to R. Thom which allows us to project all the difficulties into a space of rather small dimension.[1-4]

Thom Splitting Theorem. If $f(x)$ is a smooth function with $Df = 0$ and $Det\, D^2 f = Det\, (\partial^2 f / \partial x_i \partial x_j) = 0$ at x_o, and if q eigenvalues of the Hessian $\partial^2 f / \partial x_i \partial x_j$ are zero, then there is a smooth change of variables so that

$$f = f_{NM}(y_1, \cdots, y_q) + M^{n-q}{}_j(y_{q+1}, \cdots, y_n) \tag{4}$$

In short, there are q coordinates $y_1, \ldots, y_q$, each tangent to an eigenvector with zero eigenvalue, and a complementary set of $n-q$ coordinate directions $y_{q+1}, \cdots, y_n$, so that the function breaks up into the sum of two terms. The first is a function, f_{NM}, of the q variables $y_1, \cdots, y_q$. All second derivatives of f_{NM} vanish at x_o. The remaining function $M^{n-q}{}_j(y_{q+1}, \cdots, y_n)$ is simply a Morse saddle.

Since quadratic forms are very simple, this theorem allows us to determine what happens when the Morse Lemma is not applicable by studying a function depending on a relatively small number of variables. For example, if $n=100$ (phase space is 100 dimensional) but $q=1$, the Thom Splitting Theorem guarantees that we can understand the singularity of $f(x)$ simply by studying the singularity of the function $f_{NM}(y_1)$ of a single variable.

Thom Classification Theorem

Table 1 contains a list of all the elementary catastrophes. As seen in this table, each elementary catastrophe is the sum of two functions, one called a catastrophe germ, the other the universal perturbation. The catastrophe germ is the non-Morse function which appears in the Thom Splitting Theorem. This function describes the situation in which two or more critical points ($Df = 0$) become degenerate. The subscript k in this table identifies the degeneracy of the critical point at the origin for the catastrophe germ. The cuspoid catastrophe germ x^{k+1} describes the coalescence of k critical points at the origin.

The table of elementary catastrophes contains a list of universal perturbations in addition to the catastrophe germs. Why is such a list necessary, especially since such lists encumber neither the Implicit Function Theorem nor the Morse Lemma?

The answer is straightforward. If $Df \neq 0$ at x_o and f is perturbed: $f(x) \rightarrow g(x) = f(x) + \varepsilon p(x)$, then $Dg \neq 0$ for ε sufficiently small, the Implicit Function Theorem remains applicable, and $g = y_1$.

If $Df = 0$ but $Det\, D^2 f \neq 0$ at x_o and f is perturbed, then $Dg \neq 0$ and $Det\, D^2 g \neq 0$ for ε sufficiently small. So it might seem that the Implicit Function Theorem becomes applicable. This is not a constructive point of view. There is a point near x_o (satisfying $\varepsilon p_i + f_{ij}\, \delta x_j = 0$) at which $Dg = 0$, $Det\, D^2 g = 0$, so that the Morse Lemma is applicable and $g = M^n{}_j(y)$. The eigenvalues of $D^2 f$ and $D^2 g = D^2 f + \varepsilon$

TABLE 1. The Elementary Catastrophes

Name	Germ	Universal Perturbation
$A_{\pm k}$	$\pm x^{k+1}$	$\sum_{j=1}^{k-1} a_j x^j$
$D_{\pm k}$	$x^2 y \pm y^{k-1}$	$a_1 x + a_2 x^2 + \sum_{j=3}^{k-1} a_j y^{j-2}$
$E_{\pm 6}$	$x^3 \pm y^4$	$\sum_{j=1}^{2} a_j y^j + \sum_{3}^{5} a_j x y^{j-3}$
E_7	$x^3 + xy^3$	$\sum_{j=1}^{4} a_j y^j + \sum_{5}^{6} a_j x y^{j-5}$
E_8	$x^3 + y^5$	$\sum_{j=1}^{3} a_j y^j + \sum_{4}^{7} a_j x y^{j-4}$

D^2p have the same signs, so the stability properties (j in the Morse saddle) are the same for both f and its perturbation.

Invariance of the qualitative properties of functions which satisfy the Implicit Function Theorem or the Morse Lemma under perturbations does not extend to non-Morse functions. For example, the simple catastrophe germ, $A_2 = x^3/3$, describes a doubly degenerate critical point at $x = 0$. Under perturbation by a linear term, $f_{NM} \to x^3/3 + ax$, the double degeneracy is lifted. If $a < 0$ the two critical points separate from each other in a canonical way ($x_\sigma = \pm \sqrt{-a}$) while if $a > 0$ the two critical points disappear altogether (i.e., they hide in the complex plane). It is truly remarkable that a relatively simple function (e.g., ax for A_2) can describe the effects of an arbitrary perturbation on the degenerate critical point which is described by the catastrophe germ. More remarkable still is that the coefficients, or control parameters, in the perturbation terms (e.g., a in the perturbation for x^3) occur linearly.

Geometry of the Fold and the Cusp

If a function depends on control parameters, then as these parameters are varied, the critical points move around, coalesce and disappear or are created in bifurcations and move away from each other. Singularities occur as the control parameters are varied. If a function depends on k control parameters, then catastrophes can be encountered which depend on up to k control parameters, but no more, except when there is symmetry. Typically, functions depend on few control parameters, often just one or two. In such cases only the catastrophe A_2 (fold) or A_3 (cusp) can be encountered. These occur so often that it is worthwhile to exhibit their properties explicitly.

Four canonical properties of the fold and the cusp are exhibited in Figs. 1 and 2.

Fold: A_2 $\quad f(x;a) = x^3/3 + ax$ (5a)

Cusp: A_3 $\quad f(x;a,b) = \pm x^4/4 + ax^2/2 + bx$ (5b)

In Fig. 1a we show what the fold family of functions looks like for $a>0$, $a=0$, $a<0$. In Fig. 1b we show the location of the critical points as a function of the control parameter a. Fig. 1c shows the value of the function at the critical points, and Fig. 1d exhibits the curvature of the function, d^2f/dx^2, evaluated at the critical points. This behavior is canonical: it is exhibited by all functions in the neighborhood of the fold catastrophe.

These canonical features are reproduced for the cusp catastrophe in Fig. 2. The cusp shaped curve in Fig. 2a is a separatrix. Within this region the function $f(x;a,b)$ has three critical points: two local minima separated by a local maximum. Outside this separatrix there is only one local minimum, which is also a global minimum. On the separatrix there is a double degeneracy as the local maximum coalesces with one of the local minima. At the origin all three critical points are degenerate. Fig. 2b shows the canonical behavior of the locations of the critical

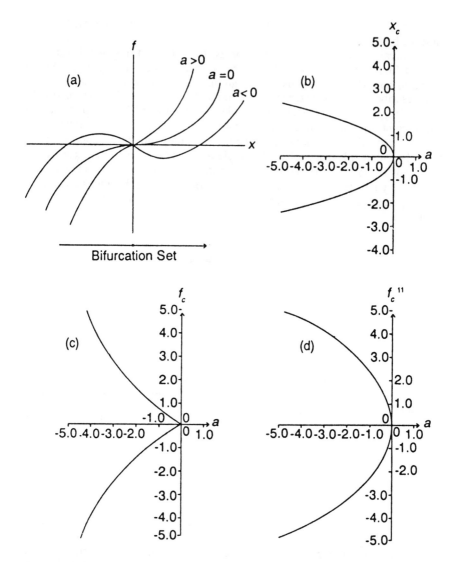

FIGURE 1. (a) Members of the fold family of functions (5a) are shown for a>0, a=0, a<0. (b) The location of the real critical points of f(x;a) are shown as a function of the control parameter a. (c) the values of $f(x_c;a)$ are shown at the critical points as a function of a. (d) The curvature $d^2f(x;a)/dx^2$ at the critical points is shown as a function of a.

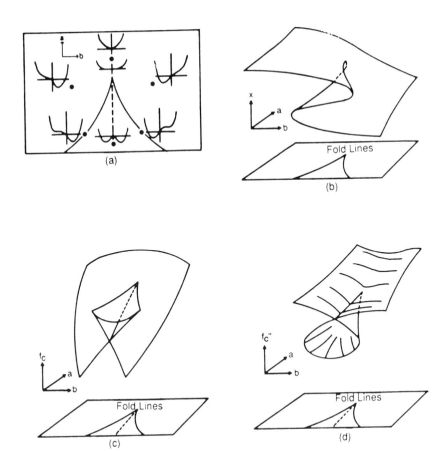

FIGURE 2. (a) Members of the cusp family of functions (5b) are shown as a function of the control parameters (a,b). (b) The two dimensional cusp catastrophe manifold $Df(x;a,b) = 0$ is shown embedded in the direct product space $R^n \times R^k = R^1 \times R^2 = R^3$. This manifold is smooth. The cusp-shaped curve in the control parameter plane is the singularity of the projection of the two dimensional manifold down onto the control parameter plane. The curves S_B, S_M are the bifurcation sets under the Delay Convention and the Maxwell Convention, discussed in Fig. 3. (c) The values of the cusp family of functions at the critical points, $f(x_c;a,b)$, are shown over the control parameter plane (a,b). (d) The critical curvature surface, $d^2f(x_c;a,b)/dx^2$, is shown over the control parameter plane.

points. The surface $Df(x;a,b) = 0$ is a smooth two-dimensional manifold in the three dimensional space $R^n \times R^k = R^2 \times R^1$, and the cusp shaped singularity in the control parameter plane exists only in the projection of the manifold $Df = 0$ onto the control plane. Figs. 2c and 2d show the values of $f(x_{cr};a,b)$ and the curvature $d^2f(x_{cr};a,b)/dx^2$ at the critical points as a function of the two control parameters a and b. The critical value surface is not a manifold because it exhibits a self intersection and sharp edges. The critical curvature surface is not a manifold because it exhibits a self intersection. These properties are canonical: they hold for all functions which exhibit cusp catastrophes.

WHY IT EXISTS

A Simple Example

We will illustrate the mathematics behind elementary catastrophe theory by means of an example. We treat a function $f(x;c)$ depending on 100 state variables, $x \in R^n$, $n=100$, and 3 control parameters, $c \in R^k$, $k=3$. At a typical point in state space $Df \neq 0$. However, there are typically isolated points (nondegenerate critical points) at which $Df = 0$. We move the origin of coordinates to one such point. At this point the constant term $f(0)$ and the first degree terms $\partial f/\partial x_i(0)$ in the Taylor series expansion of f vanish. In general, the second and higher degree terms do not vanish, and $Det(D^2f) \neq 0$. It is now possible to use the three control parameter degrees of freedom to annihilate up to three coefficients in the Taylor series of f. If terms are annihilated which do not affect the value of the Hessian, so that $Det(D^2f) \neq 0$, then the critical point at 0 remains nondegenerate and the function $f(x;c)$ is equivalent to a Morse saddle at the critical point. The Morse Lemma is not applicable only if the control parameter degrees of freedom are used to annihilate one or more second degree terms in the Taylor series expansion of f.

Two possibilities can occur with three control parameter degrees of freedom:
1. Only one eigenvalue of D^2f is zero. In this case the Thom Splitting Theorem can be invoked to produce the following decomposition

$$f(x;c) = f_{NM}(y_1;c) + M^{99}{}_j(y_2, \ldots, y_{100}) \qquad (6)$$

The three control parameter degrees of freedom can be used to annihilate the coefficients of y_1^2, y_1^3, y_1^4 in $f_{NM}(y_1)$, producing a four-fold degenerate critical point at the origin whose Taylor series description starts with the quintic y_1^5.
2. Two eigenvalues of D^2f are zero. In this case the Thom Splitting Theorem provides the decomposition

$$f(x;c) = f_{NM}(y_1,y_2) + M^{98}{}_j(y_3, \ldots, y_{100}) \qquad (7)$$

In this case the Taylor series expansion of f_{NM} begins with cubic terms. The constant term is absent by choice of origin. The linear terms are absent because the origin has been chosen as a critical point. The three quadratic terms $\partial^2 f/\partial y_1 \partial y_1$, $\partial^2 f/\partial y_1 \partial y_2 = \partial^2 f/\partial y_2 \partial y_1$, $\partial^2 f/\partial y_2 \partial y_2$ have been annihilated by using the control parameter degrees of freedom.

With three control parameter degrees of freedom it is not possible that more than two eigenvalues of D^2f can be made to vanish. The reason follows from a simple counting argument. For q eigenvalues to vanish, all $q(q+1)/2$ second degree terms $\partial^2 f_{NM}/\partial y_i \partial y_j$ in the Taylor series expansion of f_{NM} must vanish. These can be annihilated by k control parameters only if $k \geq q(q+1)/2$. For three control parameters only the two cases above are possible.

We treat the case of two vanishing eigenvalues (7) in more detail. The Taylor series expansion for $f_{NM}(y_1,y_2)$ begins with cubic terms, and all higher degree terms are generally present. It is possible to use degrees of freedom represented by a nonlinear change of variables $(y_1,y_2) \to (z_1(y), z_2(y))$, to transform away all terms of degree greater than three, and to provide a canonical form for those terms of degree three:

$$f_{NM}(y_1,y_2) = x^2 y \pm y^3 \qquad (8)$$

$(x=z_1, y=z_2)$. The control parameter degrees of freedom are no longer represented in $f_{NM}(y_1,y_2)$ as they have already been used to eliminate the three second degree terms in the Taylor series expansion. The non Morse function (8) is the catastrophe germ which represents a 4-fold degenerate critical point whose degeneracy exists in a 2-dimensional space (two vanishing eigenvalues).

What happens when this catastrophe germ is perturbed? To answer this question, we add an arbitrary perturbation $\varepsilon p(x,y)$ to the germ $x^2 y \pm y^3$. The perturbation has non zero Taylor coefficients of all degrees: $d=0,1,2,\cdots$. Since a nonlinear change of variables could previously be used to transform away all Taylor coefficients of degree four and higher, and to transform the third degree terms to the canonical germ given by (8), the same process can be applied to the perturbed function, yielding[1,2]

$$f_{NM}(x,y) + \varepsilon p(x,y) = x^2 y \pm y^3 + a_{00} + a_{10} x +$$
$$a_{01} y + a_{20} x^2 + a_{11} xy + a_{02} y^2 \qquad (9)$$

There remains the degree of freedom of shifting the origin (of ordinate (1) and coordinates (2)), eliminating three of the six remaining Taylor series coefficients. It is always possible to shift the origin to remove the terms a_{00}, $a_{20} x^2$, and $a_{11} xy$ by a rigid displacement of coordinates. As a result, the canonical form for a three parameter family of functions with a four-fold degenerate critical point having two vanishing eigenvalues is

$$f(\mathbf{x};c) = y_1^2 y_2 \pm y_2^3 + a_1 y_1 + a_2 y_2 + a_3 y_2^2 + M^{98}{}_j(y_3, \ldots, y_{100}) \qquad (10)$$

It should be noted that the three control parameters a_1, a_2, a_3 appear linearly in this perturbation.

General Procedure

The example above illustrates the general procedure for reducing families of functions depending on control parameters to a canonical form. The algorithms for constructing the germ and constructing its most general perturbation are essentially mirror images of each other.

Algorithm For Constructing Catastrophe Germs

1. Make a Taylor series expansion of the function.
2. Shift the origin of coordinates to a critical point.
3. Use the control parameter degrees of freedom to annihilate leading terms in the Taylor series expansion so that one or more eigenvalues of $D^2 f$ vanish.
4. Use a smooth nonlinear transformation to eliminate the higher degree terms in the tail of the Taylor series expansion, and to put the remaining nonzero terms in canonical form.

Algorithm for Constructing the Most General Perturbation

5. Add an arbitrary perturbation to the catastrophe germ constructed in Step 4 above.
6. Use a smooth nonlinear change of variables, as in Step 4, to eliminate the tail of the Taylor series expansion and recover the canonical form for those terms in the Taylor series expansion which constitute the catastrophe germ.
7. Shift the origin to remove as many as possible of the lower degree terms in the Taylor series expansion.

The function which remains consists of two terms (*c.f.*, (10)). One is the catastrophe germ. The other is the set of low degree terms which effect the most general possible perturbation of the degenerate critical point described by the catastrophe germ.

HOW IT WORKS

Conventions

Elementary catastrophe theory is the study of the canonical forms of potentials in the neighborhood of degenerate critical points. The potential may describe dynamics, as in the case of gradient dynamical systems (1) which describe dissipative dynamical systems, or conservative systems of the form $d^2 x_i/dt^2 = -\partial V/\partial x_i$. However, the potential itself does not specify a dynamics.

To apply the results of catastrophe theory to systems governed by a potential with slowly changing control parameters, some convention about dynamical processes must be introduced. Two conventions have been widely adopted.[1,2,4] These form extremes in a continuum which can be represented quantitatively. The two conventions are illustrated in Fig. 3.

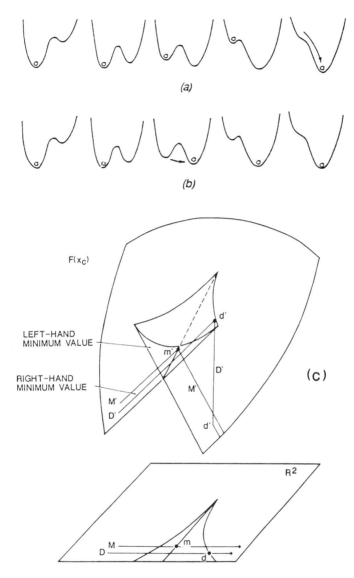

FIGURE 3. (a) Under the Delay Convention the system state remains in a local minimum until that minimum disappears; it then jumps to another accessible minimum. (b) Under the Maxwell Convention the system state is always defined by the global minimum of the potential. (c) The bifurcation sets for the two conventions are illustrated using the critical value surface (c.f., Fig. 2c) for the cusp catastrophe potential. Under the Delay and Maxwell Conventions the transition from one minimum to another occurs at m' and d' in the critical value surface, and at their projections m and d in the control parameter plane.

1. **Delay Convention.** The system remains in a local minimum as control parameters change until that minimum disappears. Then it moves to a deeper minimum.
2. **Maxwell Convention.** The system always occupies the global minimum.

The set of points (bifurcation sets) in control parameter space which describe the change of state (or phase transitions) are called bifurcation sets. When the delay convention is adopted the bifurcation set consists of the singularities in the projection of the manifold $Df(x;c) = 0$ in $R^n \times R^k$ into the control parameter space R^k. These sets can be computed by local (Taylor series) techniques, and for the cusp catastrophe A_3 are the cusp shaped curves shown in Fig. 2. For the Maxwell convention the bifurcation set can be computed by equations of Clausius-Clapeyron type.[1] For the cusp catastrophe the Maxwell set is the projection of the self intersection shown in Fig. 2c down onto the control parameter plane.

Interpolating between these two extremes, it is possible to describe the probability distribution for the occurrence of phase transitions by a Fokker-Planck equation.[1]

Catastrophe Flags

How do we know when a catastrophe is present, either in a physical system or a mathematical model? It happens that catastrophes are very obliging: they inform us of their presence by waving certain flags to attract our attention. These flags are summarized in Fig. 4.[1,2,4]

1. **Modality.** Two or more different modes of behavior exist for the same or nearby control parameter values.
2. **Sudden Jumps.** These occur when control parameters are slowly changed. As the control parameters cross bifurcation sets, the system jumps from one mode of behavior to another.
3. **Inaccessibility.** Between any two stable modes of behavior there is at least one equilibrium which is unstable. This can be identified only by very careful experimental or numerical work.
4. **Sensitivity.** For some processes the final state depends very sensitively on the initial conditions. For example, for processes leading from the single minimum to the double minimum cusp potential, the well occupied by the final system state can depend very sensitively on the starting point, for a limited number of starting points.
5. **Hysteresis.** Jumps from one mode of behavior to another do not occur for the same control parameter values as the reverse jumps under all but the Maxwell convention.

These five classical catastrophe flags are very useful for recognizing the presence of a catastrophe. However, there are many situations in which we know a catastrophe is present and even imminent, and we wish to avoid that catastrophe. Such situations could involve loss of stability of an aircraft or a ship, the collapse of a building or a bridge. Three diagnostic catastrophe flags have been developed to

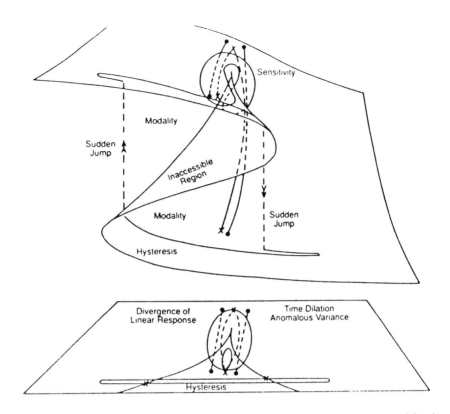

FIGURE 4. The five classical and three diagnostic catastrophe flags are illustrated for the cusp catastrophe. In a region in control parameter space with two or more stable modes of behavior, observation of any of the five catastrophe flags indicated the presence of the others. These are: multiple modality; sudden jumps from one mode to another, an unstable mode separating two stable modes; sensitivity of final state to some procedures and/pr initial conditions; hysteresis unless the Maxwell Convention is observed. Three diagnostic catastrophe flags can be observed in a single mode region, on approach to a bifurcation set. These are: divergence of the linear response function or tensor; time dilation, seen as critical slowing down or mode softening; anomalous variance of response amplitude. These can be used to locate the bifurcation set in the control parameter space and the bifurcation direction in the state variable space. Reprinted with permission from R. Gilmore, "Catastrophe Theory," in Encyclopedia of Applied Physics, Vol. 3, New York: VCH Publishers, Inc., 1991. Copyright 1992 by VCH Publishers, Inc.

prevent such surprises. In fact, they can be used to locate the singularity in state variable and control parameter space.[1]

6. **Divergence of Linear Response.** Push on something. It will move. Push a little bit harder. It will move a little bit more. The ratio of move to push is a linear response function, or tensor. As one approachs a bifurcation set (Delay Convention) the linear response function grows increasingly steep. By extrapolating this function to vertical slope it is possible to determine the location, in control parameter space, of the impending catastrophe. By constructing the corresponding eigenvector of the linear response tensor, it is possible to determine the direction in state variable space in which the bifurcation will take place.

7. **Time Dilation.** In the case of either dissipative ($k=1$) or conservative ($k=2$) gradient dynamics ($d^k x_i/dt^k = -\partial V/\partial x_i$) the approach of a singularity is associated with the vanishing of one or more eigenvalues of the linear stability matrix $\partial^2 V/\partial x_i \partial x_j$. For dissipative systems the temporal behavior is decaying exponential $e^{-\lambda t}$ (critical slowing down) while for conservative systems it is oscillatory exponential $e^{i\omega t}$ (mode softening). In either case one (or more) eigenvalues λ or frequencies ω approach zero as a singularity is approached. Further, the way in which the eigenvalue or frequency approaches zero (power law behavior) provides useful information about the singularity lurking in the background.

8. **Anomalous Variance.** In the neighborhood of a singularity, the potential becomes very flat. This means that the amplitude for oscillations, either dissipative or conservative, becomes anomalously large. The way the amplitude grows as the control parameter is varied provides information about both the location and the type of singularity. The long principal axes for anomalously large amplitudes determine the direction(s) in state space in which the bifurcation will take place.

These three diagnostic catastrophe flags are extremely useful in avoiding unexpected surprises. In particular, the presence of unexpected couplings and the sensitivity of catastrophes to these couplings can cause the locus of failure to move far from the expected range. The three diagnostics described above are then usefully used to insure that one does not unintentionally cross the bifurcation set.

An Application

To illustrate how catastrophe theory can provide useful insights, we treat here a very simple model. This consists of a column which is to be supported in a vertical position by two opposed springs situated along the x- axis and another opposed pair along the y- axis.

If one pair of springs is very strong (along the y- axis), the only failure mode will be in the x direction and the potential is[1,2]

$$V(x;F) = (1/2)(F_c - F)x^2 + (F_c/6 - F/8)x^4 \tag{11}$$

The control parameter F is the downward force applied to the top of the beam.

When the force F is less than the critical force F_c, the equilibrium displacement is $x=0$ (beam upright). As F increases to $F = F_c$, the potential exhibits an A_{+3} singularity. For $F > F_c$ (and $F < (4/3)F_c$) the equilibrium displacement is nonzero but small. Similar arguments apply if the springs in the x direction are much stronger than those in the y direction.

Collapse modes in which the equilibrium after the collapse is not too far from the pre-collapse equilibrium are called soft collapse modes.[1,2,5] They are relatively innocuous. When the post collapse equilibrium is far from the pre collapse equilibrium, the collapse mode is called hard. These are dangerous.

Cost conscious designers see little merit in using one pair of strong, expensive springs when the critical load is determined by the weaker, less expensive pair of springs. Why not design this column-spring system so that when failure occurs, it can occur in either the x- or the y- direction? This way the design cost can be minimized without reducing the failure load. Or can it?

Under such design conditions the potential governing the ideal column under the load F is[1,2]

$$V(x,y;F) = 1/2(F_c - F)(x^2 + y^2) + (F_c/6 - F/8)(x^4 + y^4) - 1/4\, F\, x^2 y^2 \quad (12)$$

In addition to the terms expected from (11) above, there is a term $(-1/4\, Fx^2y^2)$ which provides a strong coupling between the two soft failure modes. This additional term suggests that a collapse mode exists in the $x = \pm y$ plane. In this plane ($x = \pm y = z$) the potential becomes

$$V(z;F) = (F_c - F) z^2 + 2(F_c/6 - F/8) z^4 - (1/4)\, F\, z^4$$
$$= (F_c - F) z^2 + (F_c/3 - F/2) z^4 \quad (13)$$

This potential no longer represents a soft collapse mode. At the critical point $F = F_c$ the coefficient of the leading nonzero term is negative (A_{-3} !). To correctly describe this system the Taylor series of the potential should be carried out to higher order. In this case an unanticipated strong coupling between two soft collapse modes has resulted in the existence of a hard collapse mode.

But wait. The results are even worse than just suggested. It is inevitable that imperfections in fabrication of the springs and column result in a less than ideal system characterized by the potential (12) with double reflection (D_4) symmetry. The algorithm for computing universal perturbations can be used to construct the most general perturbation of (12). It is[1]

$$\text{Perturbation}(x,y) = a_{10}x + a_{01}y + a_{20}x^2 + a_{11}xy +$$
$$a_{02}y^2 + a_{21}x^2y + a_{12}xy^2 + a_{22}x^2y^2 \quad (14)$$

Of prime significance is that in the hard failure direction the perturbed potential is

$$V(z;F) = \varepsilon z + (F_c - F) z^2 + (F_c/3 - F/2) z^4 \quad (15)$$

The linear term acts to further reduce the load the column can carry before hard collapse occurs, DRASTICALLY. The failure locus has a very sharp dependence on the control parameter ε. It is a power law dependence of the form[1,2]

$$F(\varepsilon) = F_c - k|\varepsilon|^{2/3} \tag{16}$$

This means that even a relatively small imperfection will drastically reduce the carrying capacity of this column. In such cases the diagnostic catastrophe flags are invaluable in demarking the limits of 'safety' for this imperfect column with unexpectedly strong mode coupling leading to the presence of a hard collapse mode.

CONCLUSION

The general program of catastrophe theory involves a search for canonical forms for solutions of equations, and an understanding of how the qualitative properties of these solutions change as the parameters on which they depend change.[1] In particular, we would eventually like to understand how such solutions bifurcate as the parameters change. Next to nothing can be said about this program.

For the much simpler program of elementary catastrophe theory a great deal can be said. This program studies a very special class of ordinary differential equations: gradient dynamical systems. In fact, this program is even more restrictive: it seeks to determine how the critical points of potentials depend on the parameters. This program has succeeded by constructing functions which describe degenerate critical points (catastrophe germs) as well as universal perturbations. These latter are functions of minimum dimension which summarize all the distinct kinds of behavior a degenerate critical point can exhibit under an arbitrary perturbation. The control parameters which appear in the universal perturbation occur linearly.[1-4]

The program of elementary catastrophe theory cannot readily be extended to more general dynamical systems (Fig. 5). New methods need to be developed. However, all types of behavior which can be exhibited by the equilibria of potentials, or gradient dynamical systems, occur for more complex dynamical systems. For example, the fold catastrophe A_2 occurs in solutions of ordinary differential equations in two different ways. Fixed points exhibit A_2 bifurcations. So also do periodic orbits. Such bifurcations are called saddle-node bifurcations. They occur in abundance in dynamical systems which can exhibit chaotic behavior.[1,2,5]

Similar remarks hold for the cusp catastrophe A_3. Fixed points of dynamical systems can undergo a cusp bifurcation. Periodic orbits can also undergo such bifurcations - in at least two ways. A symmetric periodic orbit can undergo a symmetry-breaking bifurcation of type A_3. If the orbit is asymmetric, the A_3 bifurcation is called a period doubling bifurcation. These also occur in abundance in dynamical systems which can exhibit chaotic behavior. In addition, the Hopf bifurcation is also a cusp catastrophe in a polar coordinate system.[1,2]

One would be surprised if it were not possible to classify systems of

CATASTROPHE THEORY PYRAMID

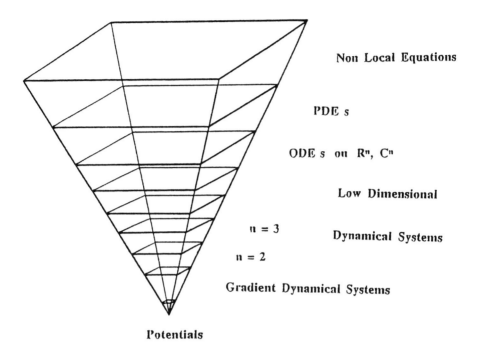

FIGURE 5. Elementary catastrophes are singularities of potentials. Potentials define the simplest types of dynamical systems: gradient dynamical systems. Their properties and behavior (modality, inaccessibility, etc.) are seen in all more complex types of systems: ordinary dynamical systems of low dimension; dynamical systems over R^n and C^n; partial differential equations; nonlocal systems of equations. As the degrees of complexity open up (vertical progression) ever more complex behavior is encountered. The elementary catastrophes exhibited by potentials with singularities form the tip of the complexity iceberg.

ordinary differential equations in part by the catastrophe which the fixed points can exhibit. For example, the Rossler system is of type A_2 since its two fixed points undergo a fold (saddle-node) catastrophe. The Lorenz system is of type A_3 since its fixed points exhibit a symmetry restricted cusp catastrophe.[1]

In addition to bifurcation behavior, the properties which catastrophes exhibit (catastrophe flags) also occur in more complex dynamical systems. Modality, sudden jumps, hysteresis, inaccessible regions, critical slowing down, anomalous variance, and sensitivity are all commonly occurring phenomena in systems of ordinary differential equations.

As illustrated in Fig. 5, elementary catastrophes associated with potentials sit at the base of all more complex dynamical behavior. They are expected to occur in systems of ordinary differential equations on R^n (and C^n), in partial differential equations, and sets of nonlocal equations. Naturally, as the equations increase in complexity, ever more complex bifurcations can be expected. The elementary catastrophes which we understand form just the tip of the iceberg.[1]

REFERENCES

1. Gilmore, R., *Catastrophe Theory for Scientists and Engineers*, New York: Wiley, 1981.
2. Gilmore, R., "Catastrophe Theory," in *Encyclopedia of Applied Physics, Vol 3*, New York: VCH Publishers, Inc., 1992, pp. 85-119.
3. Arnol'd, V. I., *Catastrophe Theory (Second Edition)*, Berlin: Springer-Verlag, 1984.
4. Poston, T., and Stewart, I. N., *Catastrophe Theory and its Applications*, London: Pitman, 1978.
5. Thompson, J. M. T., and Stewart, H. B., *Nonlinear Dynamics and Chaos*, New York: Wiley, 1986.

Robert Gilmore received his bachelors degrees in Physics and in Mathematics from MIT in 1962, and his Ph.D. in Theoretical Physics from MIT in 1967. He left the faculty of MIT in 1972 for the University of South Florida, where he remained until 1979. After a brief interval at the Institute for Defense Analyses in Washington, DC, he joined the faculty at Drexel University in Philadelphia, PA in 1981, where he is currently Professor of Physics. He is the author of two books, *Lie Groups, Lie Algebras and Some of Their Applications* (NY: Wiley, 1974; republished by NY: Dover, 1994) and *Catastrophe Theory for Scientists and Engineers* (NY: Wiley, 1981; republished by Krieger, 1994). He has worked in the fields of laser physics, atomic, molecular and nuclear physics, mathematical physics, catastrophe theory, and chaos. His current interests include analysis of large, multidimensional data sets to search for coherent structures, and the analysis of these structures using methods of nonlinear analysis.

Fractal Structures and Processes

James B. Bassingthwaighte, Daniel A. Beard,
Donald B. Percival, and Gary M. Raymond

*National Simulation Resource, Department of Bioengineering,
University of Washington, Seattle, Washington 98195*

Abstract. Fractals and chaos are closely related. Many chaotic systems have fractal features. Fractals are self-similar or self-affine structures, which means that they look much the same when magnified or reduced in scale over a reasonably large range of scales, at least two orders of magnitude and preferably more (Mandelbrot, 1983). The methods for estimating their fractal dimensions or their Hurst coefficients, which summarize the scaling relationships and their correlation structures, are going through a rapid evolutionary phase. Fractal measures can be regarded as providing a useful statistical measure of correlated random processes. They also provide a basis for analyzing recursive processes in biology such as the growth of arborizing networks in the circulatory system, airways, or glandular ducts.

OVERVIEW

Fractals, represented in structures and in processes, are ubiquitous in nature. It is intriguing that the ubiquity itself is contradictory to the notion that life is dominated by independent random processes. As we shall see, pure, uncorrelated randomness is but a special case in the fractal world. Randomness implies the absence of any correlation between one event and the next. In contrast, fractals describe objects and events which are characterized by consistencies in the degree of correlation between neighboring events or structures. The correlation may be positive or negative. Such correlations have not been completely ignored in the statistical literature, but the recognition that they by far outnumber pure randomness has until recently escaped perception, or perhaps, like the strange attractors of chaotic nonlinear dynamical systems, have been regarded somehow as "pathological" and as not worthy of a research statistician's effort. The tide has now turned.

In his introductory work Hurst (1951) examined many different phenomena. Not only did he examine river flows, but also varves (mud layer thicknesses on lake bottoms), tree rings, wheat prices, and annual rainfalls. His paper was not yet published when it was sent to Feller, who was by then already known as an outstanding statistician. In his 1951 paper Feller reflects on how forcefully Hurst's data and analyses struck him:

> "Now there is available a huge body of statistics concerning annual water levels of rivers and lakes all over the world. It has naturally been assumed that such levels could reasonably be treated as the cumulative effect of sums of random variables, but in an interesting paper H.E. Hurst discovered puzzling systematic departures. In fact, Hurst has collected an impressively large statistical material relating to water levels and other phenomena which seems to bear out the contention that the observed adjusted ranges do not increase, as expected, like the square root of the observational period T, but like a higher power T^c. The most surprising feature is the stability of the observed values of the exponent c: it varies only from 0.69 and 0.80, with a mean of 0.729 and standard deviation 0.092. Within the several separate groups of phenomena the stability of c is even greater. Hurst himself has not attempted an explanation of his interesting discovery.
>
> "It is conceivable that the phenomenon can be explained probabilistically, starting from the assumption that the variables X_k are not independent...we are here confronted with a problem which is interesting from both a statistical and a mathematical point of view."

Feller began an attempt to develop an analytical approach to predicting what Hurst needed, namely projection beyond the data set in order to predict *R/S, the range of excursions of the integral of a signal divided by the standard deviation of the signal*. He succeeded in deriving a result for only the trivial case of the Hurst exponent $H = 0.5$, which is the value for independent random fluctuations, white noise. It seemed from Feller's (1951) introduction and discussion that the revelation that many natural times series were not Gaussian or Markovian processes would soon take the statistical world by storm. Such was not to be the case. In Feller's great treatise of 1968 there is nary a mention. There has been, however, a development of a large body of knowledge on Levy processes, correlated time series, in which there is much mathematical insight, for example, as in Ossiander and Pyke (1985) and brought into modern perspective in Beran's book (1994).

Mandelbrot brought Hurst's work to the attention of scientists outside of hydrology and labelled the exponent H as the Hurst coefficient. H is a measure of "roughness" of the signal: H near 1.0 indicates a high degree of smoothness and strong positive correlation between near neighbors in the series, and H near 0 indicates a high degree of roughness and strong negative correlation. The Hurst coefficient is independent of the Euclidean dimension in which the object is embedded, and so is easier perhaps to think about than the fractal dimension, D.

The two are precisely related: $D = E + 1 - H$, where E is the Euclidean or embedding dimension.

One reason for the failure of "popular" consideration by biostatisticians and mathematical statisticians to tackle the methodologies of correlated time series more deeply has been the problem of distinguishing a stationary fractal time series from a nonstationary process. "Stationarity" in practice means that a signal has the same statistical properties at different times; a fractal Brownian signal, containing correlated runs of high values or low values, may often fail simple tests for stationarity performed on short sequences. For long-memory, positively-correlated, fractal time series, the definition of nonstationarity is not so clear, simply because long-term trends are a part of the signal, and not an indication of a fundamental change of state.

RECURSION RULES FOR GENERATING FRACTALS

Generating Fractal Geometrical Objects

Examples such as the Koch snowflake fall into the class of geometric fractals. They are simple, beautiful, and powerful. They startle us: so much diversity is captured in such simple beginnings. The power is not so much in the "beginning," but in the process of recursion. A simple act, repeated sufficiently often, creates extraordinary, often unsuspected results. Playing with recursive operations on the computer is the key to the revelation. Create your own monsters and beauties, and the insight comes free!

The application of fractal rules to simulate biological systems is a new art form. It has been better developed for botanical than zoological forms. Lindenmayer's pioneering work on plants has been extended so effectively (e.g., Prusinkiewicz and Hanan, 1989; Prusinkiewicz et al., 1990) that one has difficulty distinguishing computer-generated plants from photographs. The use of computer-generated trees as background in films attests to the degree of realism provided by the recursive application of simple rules.

Perhaps in mammalian physiology we are too hesitant in our attempts at pictorial simulation. There are no books on simulating lungs, glands, brains, or blood vessels to levels of artistry that one can enjoy. There are, however, branching stick diagrams of vascular trees, usually pictured to show the rule or the first generation or two. The remarkable thing is that models formed from recursive application of a single rule have accurately reflected the variability, or texture within an organ, of regional flow distributions (see Chapters 10 and 11 of Bassingthwaighte, Liebovitch and West, 1994). Just think how much better these algorithms might be if they used three or four rules recursively, as is done for some of the simulated plants.

The engine of creation is the recursively applied "copy machine." The simplest single-level recursion rule is

$$x_{n+1} = f(x_n), \quad (1)$$

and the commonest variant is $x_{n+1} = f(x_n) + c$, where x_{n+1} is the output of a copy machine $f(\cdot)$, meaning that x_{n+1} is a function of whatever is provided to the operator in the place of the dot; in this case the dot is replaced by x_n. The c is of great importance, and can be regarded as the influence or initial value provided by the controller for the operation. It may be "just a constant" but it can be the dominant controller as for the Mandelbrot set, $z_{n+1} = z_n^2 + c$, where z and c are complex numbers. The c is the sole controller when the initial value x_0 is zero. The processing unit or copy machine receives the previous output as one input, and an instruction from a control element C as a second input, c, for each iteration.

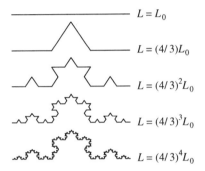

FIGURE 1. Koch curve generated by replacement of the straight line of length L_0 with a specific crooked line. The line of length L_0 is the "initiator"; the "generator" has the form of the second line, the first iteration. At each stage of the recursion each "initiator" piece is replaced by a "generator" piece. For the infinite recursion the fractal $D = \log 4/\log 3 = 1.2618$; the iterations shown are a "prefractal," and the construction is of the "dragon" type.

Recursive replacement of a line segment by a set of shorter segments is the method for creating fractal objects such as the Koch curve (Fig. 1, from Bassingthwaighte, 1988) Recursive line replacement also involves sidedness. For any generator which is symmetric about the initiator no second rule is needed. For the Koch curve, the implicit definition is that the replacement for each line segment puts the pair of line segments replacing the middle third always on the same side of the line; if the fourth line segment were replaced with a down-going generator, then the picture is quite different (Fig. 2). (Most "fractal" figures are formally "prefractals", so named because the number of generations is less than infinite. By this token nothing in biology or physics is ever fractal, since all except perhaps the universe is finite.)

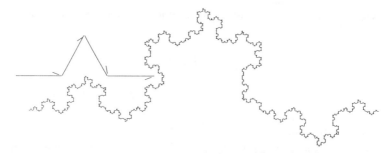

FIGURE 2. A variant on the Koch curve. The fourth line segment is replaced by the same generator, but facing toward the opposite side of the initiator line from the other three line segments. Five generations make up this "prefractal."

Generating Fractal Time Series

A variable changing with time has units of something other than time; because the units on the two axes are different, fractal time series are self-affine, but not self-similar. Most of the methods for generating fractal (or more correctly pseudo-fractal) time series use a random number generator and a series of sinusoidal terms. The reason for this is an attempt to approximate the scaling behavior found in the Fourier power spectrum where the rate of diminution in the spectral power diminishes logarithmically as a function of the frequency, the $1/f^\beta$ characteristic of fractal power law relationships. In historical order these are:

1. Weierstrass-Mandelbrot fractal function. This is a summation of sines, appropriately logarithmically scaled in amplitude and frequency, but with phases randomized. This is a continuous function that is not differentiable. See Feder (1988).
2. Weierstrass-Mandelbrot cosine function. This is the same as the Weierstrass-Mandelbrot except that there is no phase randomization (Feder, 1988).
3. Spectral Synthesis Method. Sum of sinusoidal functions having randomized phases and amplitudes, with amplitude scaling at each frequency proportional to frequency to a power. It differs from the Weierstrass function in using a finite series approximation (Saupe, 1988).
4. Fractional Gaussian Process of Davies and Harte (1987). This method, again based on Fourier spectral synthesis, uses an algorithm that gives exact performance in providing the required power law behavior and correlation. It was described in a brief appendix to their paper, and has stimulated Wood and Chan (1994) to extend the method further and to apply the approach to n-dimensional signal generation. These are probably the best ways of generating fractional Brownian noises or motions. (A Brownian motion is the integral of a Brownian noise.)

5. Successive Random Addition. Starting with a straight line, replace the middle and endpoints by using a Gaussian random displacement to create two line segments. Repeat recursively using amplitude scaled down proportionately (Saupe, 1988; Bassingthwaighte et al., 1994). This technique appears to be an approximation and needs clear evaluation. Midpoint replacement alone is not adequate.

Pure or simple fractal signals are shown in Fig. 3 for fractional Brownian noises with $H = 0.2$, 0.5 and 0.8 in the left panel. Their integrals, fractional Brownian motions, are shown in the right panel.

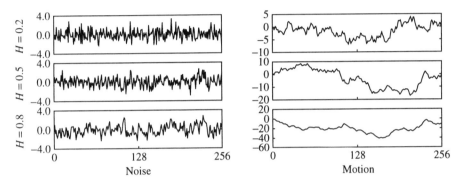

FIGURE 3. Fractional Brownian noises and motions at three H's. The noise signals (left panel) were generated using the spectral synthesis method. The motion signals (right panel) are the running sums of the noises in the left panels.

Generating Spatial Profiles

Mandelbrot, in his classic essay, "The Fractal Geometry of Nature" (1983), describes various ways of creating fractal surfaces, landscapes and coastlines. These are given extensive coverage in the chapters by Voss and by Saupe in Peitgen and Saupe's "The Science of Fractal Images" (1988). The methods include the Fourier and midpoint displacement methods described for one-dimensional signals. Feder (1988) covers these nicely also. Saupe (1988) classifies these into separate categories, noting that the displacement techniques allow increasing the resolution of the surface by increasing the number of iterations, whereas in the Fourier filtering method the resolution is determined *a priori* by the frequency range to be covered. Another method, interesting because it works surprisingly well, is to create scaled vertical displacements at cuts across a plane. This "random fault" approach (Peitgen and Saupe, 1988) uses large displacements across the fault lines at early iterations, and successively smaller displacements later.

Because of the demand for precision in characterization of surface roughness, highly precise generation techniques are now needed in order to determine the accuracy of methods of analysis. The general approach for n-dimensional signals described by Wood and Chan (1994) will be useful in this regard; its disadvantage will be that the exactness (in terms of the Fourier power spectrum and correlation properties) provided by the method requires large computer memory and severely limits the size of the surfaces or higher-dimensional objects that can be generated. The simpler techniques, like the successive random addition technique used by Voss (1985) are approximations which can be readily extended to higher dimensions, but need refinements to reach the higher accuracy required for use as reference sets. Even so, the images that Voss produced of clouds and other objects were of remarkable beauty.

Fractal Trees and Such

One can see from Fig. 1 that a single recursion rule does a lot; a minor addition augments the potential for variation in form. It is easy to see how one can use three or four rules to generate images of trees and plants (Lauwerier, 1987; Prusinkiewicz and Lindenmayer, 1990; Prusinkiewicz and Hanan, 1989) and even cellular growth (Barnsley, 1988). A host of physical phenomena are governed by only one or two rules (Vicsek, 1992; Avnir, 1989). In "Generating Fractal Time Series" and "Generating Spatial Profiles" (above) the emphasis was on the creation of random fractals whose form is governed by the Hurst coefficient and by the need to fulfill the derived expectation with respect to the Fourier power spectrum and correlation relationships. What is needed now are the developments of techniques for forging imitations of biological objects. Work on plants, particularly trees, has shown how well it can be done. A primitive example is the bush shown in Fig. 4 (from van Roy et al., 1988). Just four recursions appears to be enough to capture a realistic image. Since neurons, blood vessels, the ductal systems of kidneys, livers, pancreatic and salivary glands, and lymphatics show similar relationships in the ratios of diameters or lengths of daughter to parent segments of their branching systems these invite characterization by simple fractals or multifractals. Some beginnings will be discussed in "Biological Structures" (below).

MEASURING FRACTALS

By measuring fractals, we mean the art and science of obtaining a reasonably reliable estimate of the fractal dimension of the data. Clearly, fractal analysis methods are still primitive, and are only now being tested for their accuracy on known fractals. A major problem impeding the testing of the analysis routines has been the lack of assurance that the methods for generating "known true fractal" data sets were in fact creating true fractals. For example, our own critique of the dispersional analysis method (Bassingthwaighte and Raymond, 1995) depended on the assumption that the spectral synthesis method of Voss (1988) was accurate.

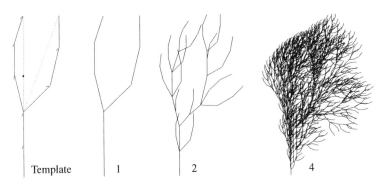

FIGURE 4. Bush, constructed from the template shown. The form of the object shown is for one, two and four iterations.

We have since discovered that this method failed, particularly badly for signals with Hurst coefficients less than 0.5, and that the method of analysis was in fact better than reported, as we will show below. This is a typical "cart and horse" problem, and where a generating method and an analysis method are both unproven, one is hard pressed to say much about either. This is why we so emphasized in "Recursion Rules for Generating Fractals" the need for generating fractals whose characteristics are known precisely.

Measuring the Fractal Dimension of Time Series

The biological processes underlying the many time series that have been characterized as "fractal" are not known, but there is no doubt that the positively correlated long memory processes are strikingly different from random processes, and that some of the bases for the correlation are being teased out by experimental work. Perhaps because of the long heritage and relatively easy instrumentation of cardiovascular, respiratory and neural signals, much of the data comes from these systems. Neural action potential intervals and spike frequencies and channel conductances have long been noted to demonstrate "$1/f$" spectra, power law spectral density functions where amplitude diminishes by a scalar fraction with each scalar multiplication of the frequency, a kind of fractal "self-similarity". Similar power law spectral densities have been observed for electrocardiographic signals (Goldberger et al., 1985; Goldberger, 1992; Yamamoto and Hughson, 1991, 1993), blood pressure records (Holstein-Rathlou and Marsh, 1994), and ventilation of the lungs (Hoop et al., 1993). Quite often, however, the power spectral density does not exhibit a simple power law decay of the form $1/f^\beta$, but does exhibit a broad peaked spectrum and slow decay in near-neighbor correlation. Although a variety of such signals were reviewed in our recent book

(Bassingthwaighte, Liebovitch and West, 1994), the past year has seen a surge in efforts to define the characteristics of such signals.

Much of the data on biological systems are time series data, measures of one or more variables obtained at evenly spaced intervals. At best only a few of the many possible measures are made. For systems of nonlinear sets of equations Takens (1981) pointed out that any one variable contains information on all of the related variables; this meant that nonlinear dynamical systems of N equations could be revealed by phase space embedding using at least N embedding dimensions, and that the structure revealed would normally have a correlation or information or fractal dimension less than N, most commonly a noninteger dimension. The more the correlation amongst the variables the lower the fractal dimension. Now in this essay we do not study nonlinear dynamical systems, which are highly predictable over the short run, but instead focus on stochastic fractal systems, which are predictable only on a statistical basis, and not on a deterministic or continuous basis. While it is correct to think of nonlinear dynamical systems in the chaotic mode as having fractal characteristics, these are usually secondary, such as the distances between the rings of Saturn or the distances between points in a Poincaré section through the attractor. Instead of looking for higher dimensional structures to be formed out of a time series, for most biological time series we look for evidence of long memory, expressed as correlation between members of the series that extends over long times. For simple fractal signals, unmarred by periodic or other interfering signals, the fractal dimension gives a direct measure of the long range correlation:

$$r_n = \frac{1}{2} \{ (n+1)^{2H} - 2n^{2H} + (n-1)^{2H} \}, \qquad (2)$$

where n is the distance between elements, calculated over the whole length of the series, r is the correlation coefficient between signal elements separated by $n-1$ elements, and H is the Hurst coefficient (e.g., in Bassingthwaighte and Beyer, 1991, and Beran, 1994).

Measures such as the caliper (or yardstick) method and the box counting method (which provides the box fractal dimension, or Minkowski-Bouligand dimension as it is called by Moreira et al., 1994, when circles are used instead of squares) have tendencies toward bias, or are otherwise not so robust (Schmittbuhl et al., 1994) when applied to one-dimensional series. Both are better for lines or surfaces or objects embedded in higher dimensions. The state of the art in time series analysis is not yet at all refined, and several methods are in vogue, even those for which there is little or no documentation with respect to accuracy, bias, and the length of a series required to obtained a defined level of accuracy. Hurst's method, though the earliest and though extensively explored by Mandelbrot and Wallis (1968, 1969) turns out to be rather weak, a disappointment to those of us who would like to see Hurst's legacy remain in his method as well as in the use of H for describing the prime characteristic of fractal signals. The methods are described briefly and commented upon in the current state of knowledge, a reminder that much remains unknown about these methods.

The relationships between the various measures of a simple fractal are given in Table 1. In theory, any of these measures gives the same insight into the nature of the signal. In practice, they may be obtained via quite different techniques; since biological signals are never pure simple fractional Brownian noises, the various measures differ not only in the estimated value, but in their degrees of reliability for signals of a given length.

TABLE 1. Fractal D, H, β and r_1 for Simple Fractional Brownian Noises (or Applied to the Successive Differences Between Elements in a Fractional Brownian Motion).

	D	H	β	r_1
$D =$	D	$2 - H$	$(3 - \beta)/2$	$\dfrac{3 - \log_2 (r_1 + 1)}{2}$
$H =$	$2 - D$	H	$(1 + \beta)/2$	$\dfrac{1 + \log_2 (r_1 + 1)}{2}$
$\beta =$	$3 - 2D$	$2H - 1$	β	$\log_2 (r_1 + 1)$
$r_1 =$	$2^{3-2D} - 1$	$2^{2H-1} - 1$	$2^{\beta} - 1$	r_1

where D is the fractal dimension, H is the Hurst coefficient, β is the exponent of the Fourier power spectral density function $1/f^{\beta}$ where f is frequency, and r_1 is the near-neighbor correlation coefficient and the first term of the autocovariance function, r_n.

First approaches to the analysis of a times series $x(t)$ or x_i, $i = 1, N$ for equal increments in time are to take simple measures, the mean, the standard deviation, the probability density function of values. At least as important as these is visual examination: Are there obvious recurring features even if not periodic? Are there trends? Are there jumps from one texture to another?

Next approaches might include calculation of the Fourier spectrum and the autocovariance functions. One reason for doing these next before embarking on analyses for fractal characteristics is to approach the question of whether or not the signal might come from a low order nonlinear dynamical system in chaotic mode. In such a case the power spectrum may show broad spectral peaks, but give no clear hint of $1/f$ behavior that characterizes simple random fractal noises. The falloff in correlation with distance between elements then gives an indication of the time lag to be used in embedding procedures for determining the correlation dimension or other dimension of the dynamical system. The analysis of chaotic systems is described elsewhere in this symposium. When $1/f$ behavior, the logarithmic decrease in amplitude with frequency, is found, then the direction to take is to consider the signal as a random fractal, a fractional Brownian noise, a fractional Brownian motion, or a more complex or composite signal.

1. Power spectral analysis: Fourier spectral analysis is based on taking the Fourier transform of the data set. When a signal has self-similar or self-affine structuring so that it appears much the same over a wide range of scales, then this is reflected in the Fourier power spectrum: |Amplitude|^2 is proportional to $1/f^\beta$. This means that the logarithm of the real part of the transform plotted against the logarithm of the frequency yields a straight line with a slope $-\beta$. If the signal is white noise (random independent events) then $\beta = 0$, i.e., the power of the component of the signal is the same at all frequencies; this random fractional Brownian noise has a Hurst coefficient of 0.5. The slope, $\beta = 2H - 1$ for fractional Brownian noises. For signals with negatively correlated nearest neighbors, H is less than 0.5, and for those with positively correlated nearest neighbors H is greater than 0.5.

2. Dispersional analysis: The method of "dispersional analysis" (Bassingthwaighte, 1988; Bassingthwaighte, King and Roger, 1989) is clearly better than R/S analysis for fractal signals with negative near-neighbor correlation, and somewhat better but still imperfect for highly correlated signals (Bassingthwaighte and Raymond, 1995). This method is based on finding the variability in the average signal over windows m elements long. The Hurst coefficient is 1 minus the slope of log variance versus log m. Figure 5 (from Bassingthwaighte et al., 1988, their Fig. 3) shows the idea for a 256 element series, showing the bin averages over a succession of longer binning lengths, m. Figure 6 shows a plot of log $RD(m)$ versus log m, where RD is relative dispersion, the standard deviation divided by the mean. (Because the mean is for the whole time series it has no effect on the slope, and can be omitted from the calculation.)

   Application of dispersional analysis to fractional Brownian noises generated using the Davies and Harte fractional Gaussian process shows that dispersional analysis is good even for short series at low H (rough processes) and fairly good, but biased toward underestimation, at high H (smooth), Fig. 7. For short series, the variance of the estimates precludes making meaningful corrections for the bias, but for longer series bias correction may be useful. The accuracy of estimates of H at varied H is shown in Fig. 7.

3. The Hurst rescaled range or R/S analysis: Our view of this frequently used technique is that it is subject to substantial systematic error in estimating the Hurst coefficient. The method (Hurst, 1951; the several Mandelbrot and Wallis papers in 1968 and 1969; Bassingthwaighte and Raymond, 1994) is based on predicting the size of a reservoir needed to even out the flows of a river subject to runs of dry or wet years. The integral of the time series (adjusted to have a zero mean) is calculated; then this derived waveform is analyzed by finding the range from the highest to the lowest value within each of many time windows of equal length τ and divided by the standard deviation within the same

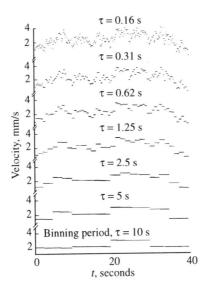

FIGURE 5. Successive averaging of pairs of elements of a one-dimensional correlated signal. The group sizes enlarge by a factor of two for each row; the group means are plotted over the range included in the group. Fractal $D = 1.38$, $H = 0.62$.

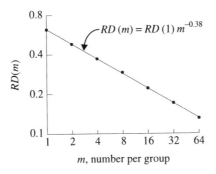

FIGURE 6. Fitting of the data of Fig. 4.2 and Table 4.1. The log of the relative dispersion (RD) of the mean of each group is plotted versus the log of the number of data values m in each group. The log-log regression is a good fit. Fractal $D = 1.38$; $RD(m = 1) = 0.623$.

window. The average value of R/S is obtained for this τ, and the same process is used to get R/S for window lengths over the possible range.

A plot of the logarithms of the average R/S at each τ versus log τ has a slope equal to H, the estimated Hurst exponent. The method is biased toward overestimation of H at low H (rough signal with negative near-neighbor correlation) and toward underestimation at high H (smoother signals with positive near-neighbor correlation). Estimates are correct near $H = 0.72$. Correcting for bias is not possible except on a statistical basis because the

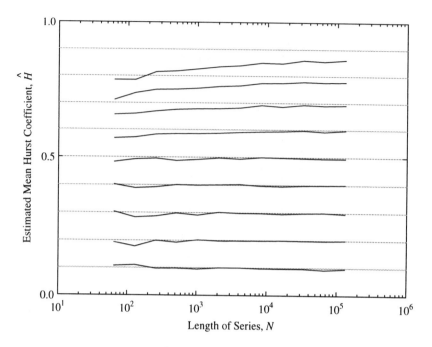

FIGURE 7. Means and SDs of $\hat{H}$, the estimated Hurst coefficient, using dispersional analysis of fractional Brownian noises generated by the Davies and Harte FGP method. The bars indicate 1 SD. Each point is the mean of 100 trials at each value of H and at each series length. There are 10,800 series represented here.

confidence limits on the estimates are broad, even for series as long as a million elements. A variant method using local trend correction is only a little better, and only for signal with $H > 0.5$ (Bassingthwaighte and Raymond, 1994).

4. Scaled windowed variance (Fano) analysis: An assessment of the scaled windowed variance or Fano Factor or variable bandwidth method (e.g., Lowen and Teich, 1995) has not been fully presented in the literature, but some evidence is provided by Schmittbuhl et al. (1994) and by Hausdorff et al. (1995), suggesting that it is quite reasonable. The method is applied to a motion signal rather than a noise. In other words if the signal is a fractional Brownian noise, then the signal is to be integrated before the analysis is begun, This decision is made from the slope β of the power spectral density obtained by Fourier analysis, as in $|A| = 1/f^{\beta}$. If $-1 < \beta < 1$, then the signal is considered to be a fractional Brownian noise. If $\beta > 1$, then the signal is considered as a fractional Brownian motion, and the scaled windowed variance method is applied directly.

5. Autocorrelation analysis: The autocovariance function can be taken from the Fourier spectral analysis or calculated directly. Correlation is the hallmark of

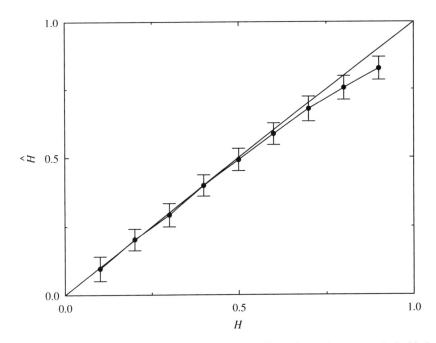

FIGURE 8. Estimates of H versus true H for fractional Brownian noises generated with the method of Davies and Harte. The bars are SDs for 100 series of 1024 points at each H.

fractal signals, a field now characterized as long-memory processes (Beran, 1994). The translation from β, D, or H to the nearest neighbor correlation coefficient is given in Table 1. The autocorrelation function, given in Eq. 1, is itself self-similar, verging into a power law relationship with a slope of $2H - 2$:

$$\frac{r_n}{r_{n-1}} = \left(\frac{n}{n-1}\right)^{2H-2}, \qquad (3)$$

where n is the index or distance between elements (Mandelbrot and Van Ness, 1968, and Bassingthwaighte and Beyer, 1991). Fitting observed autocovariance functions calculated from data sets with Eq. 1 is a good way to estimate H and D, but it appears that for single data sets one can do as well with the first few n alone and not waste time computing beyond r_5. Although calculating autocovariance is an old technique, how accurate it is for known fractal series is not thoroughly worked out (Schepers et al., 1991).

Summary of One Dimensional Series Analysis

Presumably the methods for the time series analysis of complex signals must have their basis in the methods that work for simple, pure, fractal time series. Thus our emphasis on the quality and utility of the simple methods is a kind of "back to the basics" approach. But biological time series will often be complex, meaning that they will likely be composites of more than one type of signal and may well be nonstationary. Nonstationarities in a system create a kind of "local correlation" in that when there are periods having different mean levels, this will be translated into a Hurst coefficient greater than 0.5 even when the local signals are pure white noise. An overall approach to this problem is a brute force one: prove out all the varied methods for pure signals, and then combine them in varied ways to look for methods of characterizing more complex signals. Yamamoto and Hughson (1991) have embarked on this process in attempting to separate fractal and sinusoidal contributions to a signal. Other methods are being developed and existing methods improved. Saito et al. (1994) present a large number of papers dealing with maximum entropy methods. Weigend and Gershenfeld (1994) have put together a most interesting volume on attempts to predict the future values of time series evolving from chaotic dynamical systems. We are at the beginning.

Measuring the Fractal Dimension of Lines and Surfaces and Cloud Densities

Symposium volumes covering some of the new approaches include Nonnenmacher et al. (1994) and Bunde and Havlin (1994). Good fundamental texts are those of Feder (1988) and of Peitgen et al. (1992). A new text with computer disks included is that of Russ (1995). It is a guide to a nicely menued program, including methods for generating time series and surfaces. The earlier work of Stepoway, Willis and Kane (1983) is useful.

A classic problem is the determination of the lengths of boundaries and coastlines. Although attributed to Richardson (1961), the earliest observation that a coastline exhibited self-similarity in its apparent length was that of Penck (1894) who provided data which show that the Istrian peninsula in the North Adriatic has a fractal dimension D of 1.18. Richardson made overt the relationship that $\log L(r)$ was a linear function of $\log r$, where r was the measuring stick length.

Schmittbuhl et al. (1994) have shown that Richardson's method, using calipers of varied lengths to measure line lengths, is not very accurate. The box counting method is better, at least for determining the lengths of lines on a plane, but is not very good for time series, which are self-affine, not self-similar, requiring that the boxes be rectangular, not square, and the measure is dependent on the relationship between the height of the box and the signal height.

The dispersional analysis method is well suited to the measurement of a property within an element of two-dimensional or three-dimensional space. Like the autocovariance function and the Fourier spectrum, the approach is independent of the Euclidean space. As for the one-dimensional method given for the time series analysis, the method proceeds by measuring the variance at the highest level of resolution available in the data set, then grouping small sets of near-neighbors together and finding the variance of the group means, then enlarging the size of the groupings repeatedly to obtain the variances at each size or resolution level. The slope of the regression of log (variance) versus log (size) is $E - D$ or $H - 1$, where E is the Euclidean dimension. For characterizing the variability of regional flows in the heart (Bassingthwaighte, 1988; Bassingthwaighte, King and Roger, 1989) we reported a fractal dimension D of about 1.2. This would be better reported as a Hurst coefficient of 0.8; the problem with reporting a fractal dimension of 1.2 is that the Euclidean dimension of a heart is 3, so the fractal dimension must be greater than 3, and should have been reported as 3.2. Our analysis grouped volume elements of the tissue together without regard for direction, implicitly assuming that the fractal relationships were isotropic, the same in all directions. Further analysis has not proven that the heart is anisotropic with respect to regional flows, so the assumption that direction is unimportant may be acceptable. However, there are situations where the endocardial regions, next to the ventricular cavity, have flows which are reduced compared to the epicardial regions near the surface of the heart; anisotropy needs to be considered then, not on a strict Euclidean basis, but on a geometric basis taking the shape of the heart into account, and certainly cylindrical geometry would be better than cubic.

BIOLOGICAL STRUCTURES

Parsimony by Repetition of Simple Rules

The growth of cells and organs is not governed by the genes directly. There are billions of capillaries in a heart, delivering material to hundreds of billions of cells, and similarly in other organs. But there are perhaps 100,000 genes defined in the 10^9 base pairs making up the DNA in the chromosomes. Thus the genes can only define proteins which then function in concert to act as rules for growth and development. An organ can then be developed by recursive (or continuous) application of these rules. Meinhardt (1982) shows many examples whereby the combination of rules subject to physico-chemical relationships function to direct growth. The cellular production of growth factors in response to local stimuli undoubtedly plays a role in regulating growth to defined forms, the kinds of forms described in such nice general terms by D'Arcy Thompson (1961).

Growth by branching is an algorithm central to the development of most organs. The reason for this common thread amongst tissues and organs is the fundamental need to obtain nutrients and remove waste metabolites. In glandular and excretory organs there is the additional need to provide conduit to export the products of cellular effort, urine from the kidney, bile from the liver, and so on. Like trees, the ratios of branch diameters and lengths and wall thickness tend to be constant from generation to generation, as observed by Suwa et al. (1963, 1973). The same applies to cardiac arteries (Kassab et al., 1993).

Kaandorp (1994) illustrates the power of iterated function systems to mimic the growth and form of sea sponges. Lindenmayer (1968) using "L-system" replacement rules recursively had introduced the notions. Barnsley's approach, iterated function systems or IFS, using repeated applications of equations for translation, rotation and scaling, is described in his informative book "Fractals Everywhere" (1988).

Fractal Organization of the Coronary Vasculature

Flows in the normal myocardium exhibit broad heterogeneity. Regional flows are seen to range from 1/3 of the mean flow to twice the mean flow, and the coefficient of variation is about 30%, when observations are made in tissue volumes of about 0.5% of the total heart mass. This measure of spatial heterogeneity is dependent upon the resolution, and is broader when the resolution is finer. The fractal dimension gives a scale-independent measure of the rate of increase of the coefficient of variation with an increase in resolution; in dog, baboon, sheep, and rabbit hearts the fractal D is about 1.2. Reporting of this value in individual hearts, along with the dispersion or CV at some reference level, e.g., 1 g volume elements, allows the reporting of the heterogeneity in a form that is independent of the details of the observation method or the degree of resolution used, and thereby serves as means of describing the data that can be directly compared in labs around the world. So this first use of fractals is as a statistical descriptor (Bassingthwaighte, 1988; Bassingthwaighte et al., 1989).

The finding of fractal dimensions greater than 1.0 means that the heterogeneity is finite and beyond the noise level, and the value below 1.5 means that the variation is not random, and, further, implies that there is a strong correlation between flows in neighboring regions (Bassingthwaighte and Beyer, 1991). The neighbor-to-neighbor correlation coefficient, $r(1)$, is directly measured by the fractal dimension; the relationship is $r(1) = 2^{(3-2D)} - 1$. In addition the falloff in correlation with separation distance between volume elements is also mathematically defined so that, approximately, $r(n) = r(1) \cdot n^{(2.2(H-1))}$, which asymptotically approaches $r(n) = r(3) \cdot (n/3)^{(2H-2)}$, where for generality we use the Hurst coefficient H where $H = 2 - D$ for one-dimensional measures. Such statements are based on the assumption that correlations in myocardial flows are the same in all directions, which is something that needs to be examined critically.

A partial explanation for these statistical observations of a particular correlation structure lies in the branching structure of the coronary network. Good descriptions of the spatial heterogeneity are provided by very oversimplified dichotomous branching models; this indicates that measures of heterogeneity are not very sensitive to particular forms of branching, unless the branching is characterized over several generations (van Beek et al., 1989).

What one expects on the basis of optimality principles is that there is association within each region of the myocardium between local work and energy requirements and local flow. Linking these, one can also expect that the transport capacity for critical substrates, particularly fatty acid, and the local metabolism of substrates for energy production (oxygen, fatty acids, glucose) will also be roughly proportional to local work. So far, only the association between local flows and local transport capacity has been satisfactorily demonstrated (Caldwell et al., 1994). What one also expects is that the vascular network will have grown in response to local needs and so has the capacity required to serve the "normal needs" of the tissues within its cognate region of supply.

Bassingthwaighte and Beard (1995, in press) analyzed the shapes of ^{15}O-water residue or washout curves following bolus injections into the inflow of blood-perfused rabbit hearts. Contrary to the rationale that washout curves tend to be monoexponential, these high resolution data sets extending over long times demonstrated that the washout curves were better fitted with power law functions (Fig. 8), with the power law slope of about -2 for the residue curves and about -3 for the outflow dilution curves.

On this basis one can examine the structures of the arterial network, asking the question "Is the hydrodynamic nature of blood flow through the network such that the observed heterogeneity can be explained?" Data on the nature of the network from the dimensions and lengths of large coronary vessels down to capillaries have been acquired, particularly over the past two decades (Bassingthwaighte et al., 1974; Arts et al., 1979; van Bavel, E., and J. A. Spaan, 1992; Kassab et al., 1993). The data on two pig hearts, acquired by Kassab et al., are the most complete acquired to date. Much more effort is needed in order to define the characteristics of coronary networks in a variety of species. While it is known that there is interspecies variation in specific features, e.g., arterial to arterial anastomoses, and the ability to grow collaterals, there is a dearth of quantitative, comparative data defining the differences amongst species and virtually none relating this to genetic differences, embryologic development, and the control of growth.

Thus, taking what is currently known, namely the data on the diameters of the 11 orders of branches of the pig coronary arteries, and the "connectivity" matrix which gives the probability of occurrence of daughter branches of given diameters from each of these orders, we can ask the question: "Can one construct mathematically an arterial network that matches in a statistical sense the measured

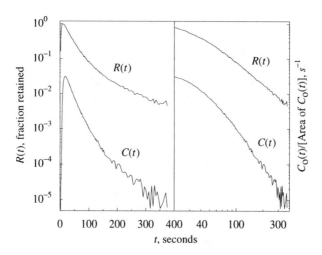

FIGURE 9. Residue curves, $R(t)$, and outflow concentration-time curves, $C(t)$, from a perfused rabbit heart after injection of ^{15}O-water into the coronary inflow. Both $R(t)$ and $C(t)$ have tails that are power law functions of time, that is, they are straight on log-log plots (*right panel*), clearly not monoexponential, being concave up on semilog plots (*left panel*).

network?" Of course, lacking data in the original data set on the three-dimensional position of each branch, or the three-dimensional or two-dimensional branching angles, or the identification of penetrating arteries, the artificial network can only be an approximation. Nevertheless, four tests of its adequacy can be formulated: 1) Does the network result in a flow heterogeneity that is appropriate (van Beek, Rogers and Bassingthwaighte, 1989)? 2) Is the spatial correlation structure in the regional flows the same as observed in real life (Bassingthwaighte and Beyer, 1991)? 3) Does the time transit time distribution through the network fit a power law relationship, as observed for both intravascular markers and a tracer water marker (Yipintsoi et al., 1970)? and 4) Does the fractional escape rate for a water tracer retained within the organ exhibit a power law slope of −1, as found for blood-perfused isolated rabbit hearts and dog hearts? (Bassingthwaighte and Beard, 1995, in press).

Further assumptions must be made if one is to formulate a complete network. How are the arteries connected to the capillary bed; what are the spatial relationships between the arteriolar entry points into the bed and the venular draining points? How many veins are there per arteriole at each of the orders or generations?

For a first approximation we avoided these more refined approaches and simply asked what resulted from the first seven orders of arterial branching, which defines the flows into regions of about 100 mg mass in pig and sheep hearts. (There are a lot of apples-and-oranges comparisons here: anatomic data are from pig hearts, physiological data are from other species.) To do this the network must be complete to the level of the capillaries at least, although connection to the venous network will have additional but rather minor influences since the major resistances are on the arterial side.

The results were obtained through three calculations: 1) the generation of the fractal network using the connectivity matrix of Kassab et al. (1993) down to the capillary level; 2) the calculation of the flows through the network from the lengths and diameters of the connected network of segments, and assuming a constant pressure at the inflow and at the capillary level; this gives a distribution of flows that is independent of the actual overall driving and sink pressures; and 3) the calculation of transit time through each pathway using either of two assumptions, plug flow or Poiseuille flow, and ignoring inertial effects at branch points.

The results were:

1. The flow heterogeneity exhibited a relative dispersion of $30 \pm 2\%$ at the seventh generation, 60–65% at the eleventh generation.
2. The fractal dimension of the flow heterogeneity found by dispersional analysis was 1.2 ± 0.1, similar to values found in the myocardium *in vivo*.
3. The near-neighbor correlation coefficient r_1 calculated by autocorrelation was $r_1 = 0.44$ and the power slope of the r_n versus n relationship was 0.42 (compared to the observed value of 0.5) which translates into a fractal dimension of 1.2 ± 0.07.
4. The washout curve deviated strikingly from monoexponential, as required, and followed straight line relationships for log residue and log outflow concentration versus log time; the power law exponents for the residue data averaged -2.0 ± 0.1 and for the outflow curves averaged -3.0 ± 0.1, values which closely match those from experimental curves such as are seen in Fig. 8.

The best fits of the power law function to the fractional escape rates gave exponents of -1.0 ± 0.47, compatible with expectations of the exponent being -1.0.

These results, tested against the expectations from four types of experimental observations, failed to demonstrate any statistical difference between the behavior of the computed network and the experimental data. Also, the artificial network demonstrated behavior that was clearly different from monoexponential, again indicating, as did the experimental data, that monoexponential washout must be regarded from both experimental and analytical perspectives to be an inadequate and inaccurate model for whole organ indicator transport functions. This says in effect that the classical Stewart-Hamilton extrapolation (Hamilton et al., 1932) of the downslope of indicator dilution curves used traditionally to obtain the areas under the curves is inaccurate and introduces a systematic underestimation of the areas. It says also that the several unimodal functions commonly used to describe indicator transport functions, the lagged normal density curve (Bassingthwaighte and Warner, 1965), the gamma variate function (Thompson et al., 1964) and the random walk (Sheppard et al., 1954) are also inadequate as descriptors for transorgan transit time distributions since all of these become monoexponential on the tails of the curves.

The conclusion from the analyses is that reconstructions of the coronary network using data from pig hearts provide predictions of flow distributions, of regional correlations in flows, of transorgan transit time distributions and fractional escape rates that fit experimental observations. These results affirm in an indirect way the accuracy of the data collection and reporting scheme used by Kassab et al. to archive their data, and illustrate that their data are of great utility to the research community outside of their own institution.

REFERENCES

Arts, T., R. T. Kruger, W. van Gerven, J. A. Labregts, and R. S. Reneman. Propagation velocity and reflection of pressure waves in the canine coronary artery. *Am. J. Physiol.* 237:H469-H474, 1979.

Avnir, D. The Fractal Approach to Heterogeneous Chemistry. In: *The Fractal Approach to Heterogeneous Chemistry: Surfaces, Colloids, Polymers*, edited by D. Avnir. New York: Wiley, 1989, pp. 199-225.

Barnsley, M. F. *Fractals Everywhere*. Boston: Academic Press, Inc., 1988, 394 pp.

Bassingthwaighte, J. B., and H. R. Warner. Indicator dispersion in the circulation. *Am. Heart J.* 69:838-841, 1965.

Bassingthwaighte, J. B., T. Yipintsoi, and R. B. Harvey. Microvasculature of the dog left ventricular myocardium. *Microvasc. Res.* 7:229-249, 1974.

Bassingthwaighte, J. B. Physiological heterogeneity: Fractals link determinism and randomness in structures and functions. *News Physiol. Sci.* 3:5-10, 1988.

Bassingthwaighte, J. B., R. B. King, J. E. Sambrook, and B. H. Van Steenwyk. Fractal analysis of blood-tissue exchange kinetics. In: *Oxygen Transport to Tissue X, Adv. Exp. Med. Biol. 222*, edited by M. Mochizuki et al. New York: Plenum Press, 1988, pp. 15-23.

Bassingthwaighte, J. B., R. B. King, and S. A. Roger. Fractal nature of regional myocardial blood flow heterogeneity. *Circ. Res.* 65:578-590, 1989.

Bassingthwaighte, J. B., C. Y. Wang, and I. S. Chan. Blood-tissue exchange via transport and transformation by endothelial cells. *Circ. Res.* 65:997-1020, 1989.

Bassingthwaighte, J. B., R. P. Beyer, J. Schepers, and J. H. G. M. van Beek. Intramyocardial flow variation: Do fractal spatial statistics reflect ontogeny. *FASEB J.* 5:A1424, 1991.

Bassingthwaighte, J. B., and R. P. Beyer. Fractal correlation in heterogeneous systems. *Physica D* 53:71-84, 1991.

Bassingthwaighte, J. B., and G. M. Raymond. Evaluating rescaled range analysis for time series. *Ann. Biomed. Eng.* 22:432-444, 1994.

Bassingthwaighte, J. B., L. S. Liebovitch, and B. J. West. *Fractal Physiology*. New York, London: Oxford University Press, 1994, 364 pp.

Bassingthwaighte, J. B., and G. M. Raymond. Evaluation of the dispersional analysis method for fractal time series. *Ann. Biomed. Eng.* 23:491-505, 1995.

Bassingthwaighte, J. B., and D. A. Beard. Fractal ^{15}O-water washout from the heart. *Circ. Res.* 77, 1995 (in press).

Beran, J. *Statistics for long-memory processes.* New York: Chapman & Hall, 1994, 315 pp.

Bunde, A., and S. Havlin. *Fractals in Science.* Berlin: Springer, 1994, 208 pp.

Caldwell, J. H., G. V. Martin, G. M. Raymond, and J. B. Bassingthwaighte. Regional myocardial flow and capillary permeability-surface area products are nearly proportional. *Am. J. Physiol.* 267 (*Heart Circ. Physiol.* 36):H654-H666, 1994.

Davies, R. B., and D. S. Harte. Tests for Hurst effect. *Biometrika* 74:95-101, 1987.

Feder, J. *Fractals.* New York: Plenum Press, 1988, 283 pp.

Feller, W. The asymptotic distribution of the range of sums of independent random variables. *Ann. Math. Stat.* 22:427-432, 1951.

Feller, W. *An Introduction to Probability Theory and Its Applications.* New York: John Wiley & Sons, Inc., 1968.

Goldberger, A. L., V. Bhargava, B. J. West, and A. J. Mandell. On a mechanism of cardiac electrical stability: The fractal hypothesis. *Biophys. J.* 48:525-528, 1985.

Goldberger, A. L. Fractal mechanisms in the electrophysiology of the heart. *IEEE Eng. Med. Biol.* 11:47-52, 1992.

Hamilton, W. F., J. W. Moore, J. M. Kinsman, and R. G. Spurling. Studies on the circulation. IV. Further analysis of the injection method, and of changes in hemodynamics under physiological and pathological conditions. *Am. J. Physiol.* 99:534-551, 1932.

Hausdorff, J. M., C. K. Peng, Z. Ladin, J. Y. Wei, and A. L. Goldberger. Is walking a random walk? Evidence for long-range correlations in stride interval of human gait. *J. Appl. Physiol.* 78:349-358, 1995.

Holstein-Rathlou, N. H., and D. J. Marsh. Renal blood flow regulation and arterial pressure fluctuations: A case study in nonlinear dynamics. *Physiol. Rev.* 74:637-681, 1994.

Hoop, B., H. Kazemi, and L. Liebovitch. Rescaled range analysis of resting respiration. *Chaos* 3:27-29, 1993.

Hurst, H. E. Long-term storage capacity of reservoirs. *Trans. Amer. Soc. Civ. Engrs.* 116:770-808, 1951.

Kaandorp, J. A. *Fractal Modelling: Growth and Form in Biology.* Berlin: Springer-Verlag, 1994, 208 pp.

Kassab, G. S., C. A. Rider, N. J. Tang, and Y. B. Fung. Morphometry of pig coronary arterial trees. *Am. J. Physiol. (Heart Circ. Physiol.)* 265:H350-H365, 1993.

Kassab, G. S., K. Imoto, F. C. White, C. A. Rider, Y. C. B. Fung, and C. M. Bloor. Coronary arterial tree remodeling in right ventricular hypertrophy. *Am. J. Physiol. (Heart Circ. Physiol.)* 265:H366-H375, 1993.

Lauwerier, H. *Fractals. Meetjundige figuren in eindeloze herhaling.* Amsterdam: Aramith Uitgevers, 1987.

Lindenmayer, A. Mathematical models for cellular interactions in development. I. Filaments with one-sided inputs. *J. Theoret. Biol.* 18:280-299, 1968.

Lindenmayer, A. Mathematical models for cellular interactions in development. II. Simple and branching filaments with two-sided inputs. *J. Theoret. Biol.* 18:300-315, 1968.

Lowen, S. B., and M. C. Teich. Estimation and simulation of fractal stochastic point processes. *Fractals* 3:183-210, 1995.

Mandelbrot, B. B., and J. R. Wallis. Noah, Joseph, and operational hydrology. *Water Resour. Res.* 4:909-918, 1968.

Mandelbrot, B. B., and J. W. Van Ness. Fractional brownian motions, fractional noises and applications. *SIAM Rev.* 10:422-437, 1968.

Mandelbrot, B. B., and J. R. Wallis. Computer experiments with fractional Gaussian noises. Part 1, averages and variances. *Water Resour. Res.* 5:228-241, 1969.

Mandelbrot, B. B., and J. R. Wallis. Computer experiments with fractional Gaussian noises. Part 2, rescaled ranges and spectra. *Water Resour. Res.* 5:242-259, 1969.

Mandelbrot, B. B., and J. R. Wallis. Computer experiments with fractional Gaussian noises. Part 3, mathematical appendix. *Water Resour. Res.* 5:260-267, 1969.

Mandelbrot, B. B., and J. R. Wallis. Some long-run properties of geophysical records. *Water Resour. Res.* 5:321-340, 1969.

Mandelbrot, B. B., and J. R. Wallis. Robustness of the rescaled range R/S in the measurement of noncyclic long run statistical dependence. *Water Resour. Res.* 5:967-988, 1969.

Mandelbrot, B. B. *The Fractal Geometry of Nature.* San Francisco: W.H. Freeman and Co., 1983, 468 pp.

Meinhardt, H. *Models of Biological Pattern Formation.* New York: Academic Press, 1982.

Moreira, J. G., J. Kamphorst Leal da Silva, and S. Oliffson Kamphorst. On the fractal dimension of self-affine profiles. *Phys. A Math. Gen.* 27:8079-8089, 1994.

Nonnenmacher, T. F., G. A. Losa, and E. R. Weibel. *Fractals in Biology and Medicine.* Basel: Birkhäuser, 1994, 397 pp.

Ossiander, M., and R. Pyke. Lévy's Brownian motion as a set-indexed process and a related central limit theorem. *Stochast. Processes Applic.* 21:133-145, 1985.

Peitgen, H. O., and D. Saupe. *The Science of Fractal Images*. New York: Springer-Verlag, 1988, 312 pp.
Peitgen, H. O., H. Jürgens, and D. Saupe. *Chaos and Fractals: New Frontiers of Science*. New York: Springer-Verlag, 1992, 984 pp.
Penck, A. *Morphologie der Erdoberfläche*. Stuttgart, 1894.
Prusinkiewicz, P., and J. Hanan. *Lecture Notes in Biomathematics: Lindenmayer Systems, Fractals, and Plants*. New York: Springer-Verlag, 1989, 120 pp.
Prusinkiewicz, P., A. Lindenmayer, J. S. Hanan, F. D. Fracchia, D. R. Fowler, M. J. M. de Boer, and L. Mercer. *The Algorithmic Beauty of Plants*. New York: Springer-Verlag, 1990.
Richardson, L. F. The problem of contiguity: an appendix to *Statistics of Deadly Quarrels. Gen. Sys.* 6:139-187, 1961.
Roger, S. A., J. H. G. M. van Beek, R. B. King, and J. B. Bassingthwaighte. Microvascular unit sizes govern fractal myocardial blood flow distributions. *Microcirculation* 00:000-000, 1995.
Russ, J. C. *Fractal Surfaces*. New York: Plenum Press, 1994, 309 pp.
Saito, K., A. Koyama, K. Yoneyama, Y. Sawada, and N. Ohtomo. *A recent advance in time series analysis by maximum entropy method. Application to medical and biological sciences*. Sapporo, Japan: Hokkaaido University Press, 1994, 397 pp.
Saupe, D. Algorithms for random fractals. In: *The Science of Fractal Images*, edited by H. O. Peitgen and D. Saupe. New York: Springer-Verlag, 1988, pp. 71-136.
Schmittbuhl, J., J. P. Vilotte, and S. Roux. Reliability of self-affine measurements. *Phys. Rev. E* 51:131-147, 1995.
Sheppard, C. W. Mathematical consideration of indicator-dilution techniques. *Minn. Med.* 37:93-104, 1954.
Stepoway, S. L., D. L. Wells, and G. R. Kane. *An architecture for efficient generation of fractal surfaces*, 1983, 261 pp.
Suwa, N., T. Niwa, H. Fukasawa, and Y. Sasaki. Estimation of intravascular blood pressure gradient by mathematical analysis of arterial casts. *Tohoku J. Exp. Med.* 79:168-198, 1963.
Suwa, N., and T. Takahashi. *Morphological and Morphometrical Analysis of Circulation in Hypertension and Ischemic Kidney*. Munich: Urban & Schwarzenberg, 1971.
Takens, F. Detecting strange attractors in turbulence. In: *Dynamical systems and turbulence, Warwick 1980*, edited by D. A. Rand and L. S. Young. New York: Springer-Verlag, 1981.
Thompson, D. A. W. *On Growth and Form*. Cambridge: Cambridge University Press, 1961, 346 pp.
Thompson, H. K., C. F. Starmer, R. E. Whalen, and H. D. McIntosh. Indicator transit time considered as a gamma variate. *Circ. Res.* 14:502-515, 1964.

van Bavel, E., and J. A. Spaan. Branching patterns in the porcine coronary arterial tree. Estimation of flow heterogeneity. *Circ. Res.* 71:1200-1212, 1992.

van Beek, J. H. G. M., S. A. Roger, and J. B. Bassingthwaighte. Regional myocardial flow heterogeneity explained with fractal networks. *Am. J. Physiol.* 257 (*Heart Circ. Physiol.* 26):H1670-H1680, 1989.

Van Roy, P., L. Garcia, and B. Wahl. *Designer Fractal. Mathematics for the 21st Century.* Santa Cruz, California: Dynamic Software, 1988.

Vicsek, T. *Fractal Growth Phenomena, Second Edition.* Singapore: World Scientific, 1992, 355 pp.

Voss, R. F. Random fractal forgeries. In: *Fundamental Algorithms in Computer Graphics*, edited by R. A. Earnshaw. Berlin: Springer-Verlag, 1985, pp. 805-835.

Voss, R. F. Fractals in nature: From characterization to simulation. In: *The Science of Fractal Images*, edited by H. O. Peitgen and D. Saupe. New York: Springer-Verlag, 1988, pp. 21-70.

Weigend, A. S., and N. A. Gershenfeld. *Time Series Prediction: Forecasting the Future and Understanding the Past.* Reading, MA: Addison-Wesley, 1994, 643 pp.

Wood, A. T. A., and G. Chan. Simulation of stationary Gaussian processes in $[0,1]^d$. *J. Computation. Graphical Stat.* 3:409-432, 1994.

Yamamoto, Y., and R. L. Hughson. Coarse-graining spectral analysis: New method for studying heart rate variability. *J. Appl. Physiol.* 71:1143-1150, 1991.

Yamamoto, Y., and R. L. Hughson. Extracting fractal components from time series. *Physica D* 68:250-264, 1993.

Yipintsoi, T., and J. B. Bassingthwaighte. Circulatory transport of iodoantipyrine and water in the isolated dog heart. *Circ. Res.* 27:461-477, 1970.

James Bassingthwaighte, Professor of Bioengineering, Biomathematics and Radiology at the University of Washington, received his B.A. and M.D. degrees from the University of Toronto. Following post-graduate work in Medicine and Cardiology at the Postgraduate Medical School of London, he obtained his Ph.D. in Physiology from the Mayo Graduate School of Medicine. He joined the Mayo faculty, advancing to Professor of Physiology and of Medicine. He came to the University of Washington in 1975, serving as Director of the Center for Bioengineering for 5 years. He is Affiliate Professor of Physiology at Limburg University in Maastricht, The Netherlands. Dr. Bassingthwaighte is the recipient of numerous awards such as an NIH Research Career Development Award, the Louis and Arthur Lucian Award of McGill University, the Alza Award of the Biomedical Engineering Society, the Burlington Resources Foundation Faculty Achievement Award for Research, and the Landis Award of the Microcirculatory Society, 1995. He is currently the Editor of the Annals of Biomedical Engineering and Chairman of the Commission on Bioengineering in Physiology, International Union of Physiological Sciences. He is the co-author with L. S. Liebovitch and Bruce J. West of the book *Fractal Physiology*, Oxford Press, 1994. His current research is centered on the mechanisms of flow, transport, and metabolism of substrates and hormones in the heart and throughout the body. The work emphasizes the use of quantitative mathematical models for integrative systems analysis in physiology and in image analysis; he serves as Director of the National Simulation Resource for Circulatory Mass Transport and Exchange.

Chaos, Dynamical Structure and Climate Variability

H.Bruce Stewart
Department of Applied Science
Brookhaven National Laboratory
Upton, New York 11973

Abstract. Deterministic chaos in dynamical systems offers a new paradigm for understanding irregular fluctuations. Techniques for identifying deterministic chaos from observed data, without recourse to mathematical models, are being developed. Powerful methods exist for reconstructing multidimensional phase space from an observed time series of a single scalar variable; these methods are invaluable when only a single scalar record of the dynamics is available. However in some applications multiple concurrent time series may be available for consideration as phase space coordinates.

Here we propose some basic analytical tools for such multichannel time series data, and illustrate them by applications to a simple synthetic model of chaos, to a low-order model of atmospheric circulation, and to two high-resolution paleoclimate proxy data series.

This work was supported by the CHAMMP initiative of the Office of Health and Environmental Research, U.S. Department of Energy under Contract No. DE-AC02-76CH00016.

THE NOTION OF DYNAMICAL STRUCTURE

The chaos paradigm of dynamical systems theory raises the possibility that some of the irregular oscillations observed in the laboratory and in the natural world may be explainable as the workings of deterministic dynamics. Indeed, the deterministic rules governing the behavior of even large and complex natural systems may in some instances be expressed in terms of just a handful of active modes of oscillation, which can be fully described using a small number of state variables.

A deterministic dynamical system comprises two essential ingredients: first, a state space or phase space whose coordinates are the state variables $x_1, x_2, ..., x_n$; these variables describe the state of the system at any instant of time, and the functions $x_1(t), x_2(t), ..., x_n(t)$ describe the system evolution over time. Second, a dynamical system possesses a dynamical rule which specifies completely and unambiguously for each state $X = \{x_1, x_2, ..., x_n\}$ the immediate future trend of evolution; that is, the rule uses $X(t)$ at time t to determine X a short time interval into the future [1, 2].

In many systems, there is a natural discrete unit of time, such as a day or a year. The evolution is described by a sequence X_i with index i indicating time, that is, $X_i = X(t_i)$. The dynamical rule is then conveniently expressed as an iterated function

$$X_{i+1} = F(X_i) \qquad (1)$$

where F is a vector function with vector arguments. The evolution begins from an appropriate initial condition

$$X_0 = \{x_1(t_0), x_2(t_0), ..., x_n(t_0)\} \qquad (2)$$

In other systems, it is natural to consider time flowing as a continuum, so that the dynamical rule is a differential equation

$$dX/dt = \dot{X} = F(X)$$

that is,

$$\begin{aligned} \dot{x}_1 &= f_1(x_1, x_2, ..., x_n) \\ \dot{x}_2 &= f_2(x_1, x_2, ..., x_n) \\ &\quad ... \\ \dot{x}_n &= f_n(x_1, x_2, ..., x_n) \end{aligned} \qquad (3)$$

The evolution of the system from an initial state

$$X_0 = \{x_1(t=0), x_2(t=0), ..., x_n(t=0)\} \qquad (4)$$

forward in time is the solution $X(t) = \{x_1(t), x_2(t), ..., x_n(t)\}$ of the initial value problem (3) and (4). The evolution from an initial state – a point in phase space – traces out a smooth trajectory in phase space, provided the functions $f_1, f_2, ..., f_n$ are continuous functions. That is, the magnitude of the difference $|F(X_1) - F(X_2)|$ should be small whenever $|X_1 - X_2|$ is sufficiently small. Here $|\cdot|$ indicates a distance, for example Euclidean distance, in phase space.

In discrete time systems (1) the trajectory is usually not smooth, but F should still be continuous in X.

In physical problems, such as mechanical or electrical systems, an appropriate phase space is usually apparent from the form of the laws of motion. For example, mechanical problems require a position and a velocity for each mechanical degree of freedom [1]. However, in other cases it may be a difficult task to choose an economical set of state variables, that is, a reasonably small number of coordinates which still retain the essential property of the dynamical rule: that knowing the instantaneous state $X = \{x_1, x_2, ..., x_n\}$ is sufficient to specify the immediate future trend of evolution in a completely deterministic way and without ambiguity. In spite of this difficulty, recent successes like the characterization of the historical fluctuations in the level of the Great Salt Lake using a four-dimensional state space [3, 4] show that it is possible. More particularly, it is possible to determine an appropriate state space or phase space by analyzing an observed time series, without recourse to a model based on physical laws.

Low-dimensional long-term behavior can occur in complex natural systems whose phase space would seem to be of very high dimension. Dissipation, which exists in most natural systems, causes volumes of ensembles in phase space to contract as time advances; this is the equivalent of Liouville's theorem for energy-conserving systems [2, p. 221]. In many cases, dissipation acts even more strongly, reducing the long-term fluctuations to a subset of dimension much smaller that the number of phase space dimensions suggested by the laws of motion. This is not a theorem, but a commonly observed phenomenon.

If the long-term dynamics of a system has a low-dimensional description, then one may hope that a moderately long observed trajectory will come near to every state possible for the long-term dynamics; that is, it comes near every point in the attractor. If the dynamical rule $F(X)$ is a continuous function, then it will be possible to make good short-term forecasts by identifying dynamical analogs in past observed behavior [5].

Most of the methods for detecting dynamical structure begin with the modest assumption that only one time series of a single state variable has been recorded. From a single time series, additional phase space coordinates

can be reconstructed using a procedure known as time-delay embedding [6, 7].
An excellent review of these methods has recently appeared [8]. Here we
consider situations in which two or more concurrent time series of different
variables are available for analysis. After describing the basic diagnostic of
trajectory divergence, we discuss three applications: a prototype of chaos in a
system with three-dimensional phase space; a low-order model of atmospheric
circulation with 27 phase space coordinates; and an example involving high-
resolution paleoclimate proxy data.

Identifying Dynamical Structure from Time Series

The hallmark of chaos is that evolutions from two nearby states in phase
space will gradually diverge from each other as time progresses. In mathematical terms, the system is sensitive to initial conditions. An error or
perturbation introduced at any time will grow over time, typically at a geometric rate; this makes long-term forecasting impossible. Given a moderately
long time series of all n phase space coordinates, this gradual divergence can
be verified by finding good dynamical analogies, that is, pairs of widely separated times in the observed record when the two system states were near each
other in phase space. These good dynamical analogies are manifestations of
recurrence.

Let us consider data $X(t_i)$ from a continuous time evolution sampled
discretely at equally spaced times $t = t_i, i = 1, 2, ..., N$. For each t_i, the best
dynamical analogy for $X(t_i)$ involves its nearest neighbor in phase space.
Let us say that this nearest neighbor occurs at time $t = t_{\mathcal{N}(i)}$; in determining
$\mathcal{N}(i)$ we exclude times near t_i so that the analogy belongs to distinct parts
of the trajectory and represents a true recurrence.

For each such analogy, the rate of divergence over j steps forward in time
can be measured in terms of the local spreading ratio

$$S(i,j) = |X(t_{\mathcal{N}(i)+j}) - X(t_{i+j})|/|X(t_{\mathcal{N}(i)}) - X(t_i)|. \tag{5}$$

An important mathematical fact about this spreading ratio is that (roughly
speaking) for large values of j the equivalent rate $(1/j)\ln S(i,j)$ tends to a
limit which is independent of i, and independent of the particular choice of
coordinates; in other words, the limit is an invariant quantity. The limiting
value is called the largest Lyapunov exponent, denoted λ_1. Sometimes base
2 logarithms are used, so that λ_1 is an inverse doubling time for uncertainties
or perturbations. There is in fact a spectrum of limiting rates or Lyapunov
exponents; only the largest, λ_1, is manifested in the long-term spreading of
two typical nearby trajectories. The criterion for chaos is $\lambda_1 > 0$ [8].

In the mathematical definition of Lyapunov exponents, it is assumed that the initial separation at $j = 0$ is infinitesimal, so that even for large j the separation is not too large. When dealing with a finite sample of observed data, this is of course not true, so it may not be practical to consider j large enough to obtain a true invariant quantity. Instead, one may examine the local divergence rate

$$s(i,j) = (1/j) \ln S(i,j) \tag{6}$$

and its average

$$\bar{s}(j) = (1/i) \sum_i s(i,j) \tag{7}$$

Although these are not invariant quantities, and do depend on the choice of coordinates, it is still possible to obtain from them useful information about possible dynamical structure.

A PROTOTYPE EXAMPLE

To illustrate how this can be accomplished, we first consider synthetic data generated by numerical solution of a simple system of three first-order ordinary differential equations

$$\begin{aligned} \dot{x} &= -y - z \\ \dot{y} &= x + 0.36y \\ \dot{z} &= 0.4 + z(x - 4.5) \end{aligned} \tag{8}$$

This system was devised by Roessler [9] to give an example of the simplest possible chaotic attractor, the folded band. Any trajectory of this system will, after an initial transient, settle onto a coherent three-dimensional structure. Within this coherent structure, nearby states exhibit gradual divergence over time. A typical trajectory on this attractor is illustrated in Figure 1, which shows the coordinates x, y, and z plotted as three time series above, and in phase portraits. The upper left phase portrait shows an orthogonal projection of the three coordinates with the z-axis tilted at 45 degrees, while the lower left shows a projection along the z-axis onto the (x,y) plane. On the right is a different trajectory to be discussed below.

On the average, separations are roughly doubled for each circuit around this attractor; sampling at about 60 discrete time steps per circuit, we expect $\lambda_1(j)$ to be about $(1/60)\ln 2$, or roughly 0.011.

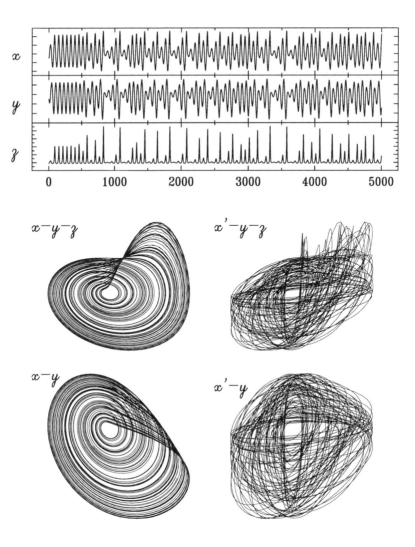

Figure 1: Time series and phase space projections of a trajectory of Roessler's equations, and phase projections of a trajectory obtained by substituting surrogate channel data x' for x.

Suppose we are given a three-channel time series $x(t_i), y(t_i), z(t_i)$; we wish to determine, from the data themselves and without knowing their origin in eqs. (8), whether they were generated by a deterministic rule. One method of diagnosis consists of computing the average local divergence rate $\bar{s}(j)$, and comparing with the divergence rate computed from a surrogate data set. By surrogate, we mean data which *from their appearance as time series* could plausibly have come from the same source, but have in fact been arranged or manipulated so that they lack dynamical structure.

Surrogate data sets have already been used for this type of diagnosis with time-delay embeddings, where only a single variable has been observed and recorded; see for example [10, 11]. Here we are considering multichannel data, so it makes sense to look at surrogate data in which only one channel has been replaced by a plausible substitute, with other channels unchanged. We then speak of a surrogate channel of data.

A simple method of generating surrogate channels is to divide the original multichannel data set into two halves. We denote the first half of the data, $i = 1, 2, ..., N/2$ by $x(t_i), y(t_i), z(t_i)$, and the second half, for $i = N/2, N/2 + 1, ..., N$, is displaced in time to $i = 1, 2, ..., N/2$ and denoted by $x'(t_i), y'(t_i), z'(t_i)$. Now x' has dynamical structure when taken with y' and z', but its structure is, time step for time step, unrelated to the structure of y and z. Nearest neighbors identified using coordinates $x'(t_i), y(t_i), z(t_i)$ will not be true dynamical analogies, and can be expected to diverge rapidly. Thus the value of $\bar{s}(j)$ for the surrogate data $x'(t_i), y(t_i), z(t_i)$ will be much larger than for the original.

On the other hand, if the original data $x(t_i), y(t_i), z(t_i)$ came not from a dynamical system but from random behavior, then the spreading rate should be large for both the original and the surrogate data. Since we do not know *a priori* what is a large spreading rate, the comparison with surrogate data is essential.

Examples of average local divergence rates $\bar{s}(j)$ for true trajectories of eqs.(8) and with surrogate channels are plotted in Figure 2. Two disjoint segments each 5000 steps in length were extracted from a longer trajectory. The two segments were spliced into a single six-channel data set to facilitate surrogate substitutions. Various embeddings and surrogate substitutions were tried, with two cases each to give a crude estimate of the variance due to finite sampling of the attractor.

The complete natural embedding of the first segment with coordinates $x(t_i), y(t_i), z(t_i)$, and the complete natural embedding of the second segment with coordinates $x'(t_i), y'(t_i), z'(t_i)$ are represented in Fig. 2(a) by the solid curves which give the average local divergence rate $\bar{s}(j)$ as a function of j, the number of steps ahead. These average rates range up to 0.020, which is of

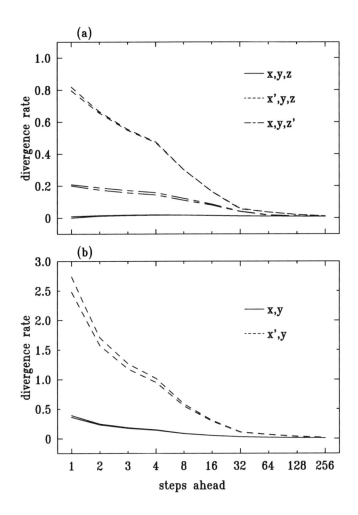

Figure 2: Average local divergence rate $\bar{s}(j)$ for trial embeddings of the Roessler band attractor in two and three dimensions; 5000-point trajectories sampled 60x per turn.

the order of the expected long-term value of 0.011. The difference between the two samples is just visible for $j = 1$. For increasing j, this difference decreases. In both cases, $\bar{s}(j)$ increases to a maximum near $j = 4$ and then decreases to the expected long-term value by $j = 64$.

The results of surrogate substitution in the x coordinate are shown in Fig. 2(a) by the evenly broken lines. In one case, $x'(t_i)$ was substituted for $x(t_i)$ in the first trajectory segment, while in the other case the opposite was done. Since the trajectories $x'(t_i), y(t_i), z(t_i)$ and $x(t_i), y'(t_i), z'(t_i)$ are of course not trajectories of a dynamical system, it is an abuse of terminology to speak of local Lyapunov exponents. Nevertheless, computing the average local divergence rates as before, we find a large increase in magnitude. If we were presented with data of unknown origin, such an effect would be evidence for the significance of the x coordinate in the dynamical structure of the trajectory.

Note that the effect of the surrogate channel upon the local divergence rates is greatest for $j = 1$, and becomes less pronounced as j increases. When trajectory self-crossing inconsistent with dynamical structure occurs, the largest separations occur in the near term. Thus, when using surrogates to test for dynamical structure, the short-term divergence rates provide better diagnosis than the long-term rates, even though the short-term rates are not invariant quantities.

The remaining two cases in Fig. 2(a), shown as unevenly broken lines, are again obtained by surrogate substitution, this time for the z coordinate. Again the importance of this coordinate in the dynamical structure is confirmed, although its significance is not so strong as that of the x coordinate.

Figure 2(b) relates to the detection of dynamical structure using an incomplete set of phase space coordinates. Only x and y are used as trial phase space coordinates; we therefore do not expect the divergences, even for large j, to approach the true Lyapunov exponent λ_1 in the three-dimensional phase space. Nevertheless, when a surrogate channel is substituted (broken curves) for the x coordinate in the partial x, y embedding, the effect on local divergence rates is unmistakable.

This suggests that it may be possible, using surrogate channel substitution, to detect dynamical structure in multichannel time series data, even if there are not enough data channels to fully embed the attractor, that is, there are not enough phase space coordinates to correctly identify good dynamical analogies.

Of course this prototype example of chaos only suggests what may happen with more complicated chaotic attractors in higher-dimensional phase space. We now turn to a more challenging example.

AN ATMOSPHERIC CIRCULATION MODEL

We next apply similar techniques to a system of 27 ordinary differential equations introduced by Lorenz as a low-order model of atmospheric circulation [12]. The model is based on a two-layer quasi-geostrophic beta-plane approximation of circulation at middle latitudes, with moisture in the atmosphere and a shallow, non-circulating ocean. The principal unknowns are the atmospheric temperature T, stream function ψ, moisture content W, and sea surface temperature S; each field is represented horizontally by an expansion in seven orthogonal spatial modes numbered 0 through 6. A global constraint eliminates the zeroth mode ψ_0 of the stream function, leaving $4 \times 7 - 1 = 27$ unknowns. Lorenz found plausible values of the parameters for which the model exhibits sustained chaotic dynamics with nearby trajectories separating exponentially in time. Similar solutions exist over a range of parameter values [13].

This model has attracted interest because its long-term behavior is suggestive when interpreted as a toy model of natural climate variability. With no daily or yearly variation in solar forcing, and numerical solution advancing four time steps per nominal day, the model has oscillations with an average period of about two weeks. Figure 3 shows atmospheric temperature at a fixed location on this time scale. A second trajectory perturbed from the first by a very small amount at time zero diverges from the first on a similar time scale. And yet the global mean atmospheric temperature – the zeroth mode T_0 – fluctuates substantially on very much longer time scales. Figure 4 shows T_0 averaged yearly (i.e. over 360-day periods) for a thousand years.

The question has therefore been raised whether the long-term fluctuations in Fig. 4 correspond to an attractor, that is, whether in some appropriate phase space there is a vector field which determines the long-term changes of mean temperature. Any such vector field would not be the one given directly by the 27 ordinary differential equations, but would presumably emerge from them in some unspecified manner. This hypothetical dynamical structure on long time scales need not exist at all, but there is some evidence for its existence, including a red spectrum, dimension estimates [14], and an argument based on rescaled range analysis and the Hurst exponent [15].

Even on the faster time scale where deterministic dynamical structure certainly governs the dynamics, the Lorenz 27-variable model has considerable interest as a dynamical system in its own right, and poses considerable challenges to the analyst. The form of the ordinary differential equations is complicated, including algebraic constraints and highly nonlinear thermodynamic relations, and the numerical solution procedure combines the modal representation to resolve some terms and conversion to a finite spatial grid to resolve other terms. These complications make it unlikely that much can

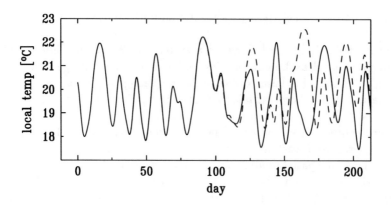

Figure 3: Two solutions of the Lorenz 27-variable moist atmospheric circulation model diverging after a small perturbation at day zero.

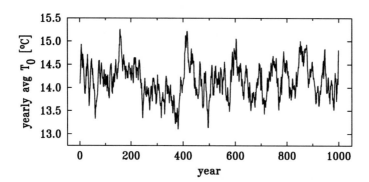

Figure 4: Long-term fluctuations of the yearly averaged global mean atmospheric temperature in the Lorenz model.

be learned by analysis of the equations. For example, at least one unstable equilibrium solution probably exists in the chaotic regime, but there is no obvious way to locate it, let alone perform a local linear stability analysis. In short, the only available tools of analysis are those which treat the numerically generated solutions as if they were produced by a black box. On the brighter side, any tools which work for this system may also be useful for the analysis of observed experimental data, even when a mathematical model of the system is unavailable.

We shall undertake two different analyses of the Lorenz 27-variable model, treating the model output as time series data. First we attempt to describe the chaotic attractor on the fast time scale, interpreted as a toy model of weather. We know there is dynamical structure, and aim to describe it efficiently by making use of multiple channels of time series data, the 27 modal coefficient time series. We use estimates of local divergence rates and comparison with surrogate channel data to find a subset of the 27 channels which gives a faithful embedding of the chaotic attractor. The goal is to identify dynamical structure in a challenging test case. In the process we will obtain an unexpected bonus of useful information.

Second we apply similar methods to yearly averaged output of the 27-variable model, interpreted as a toy model of climate. Here we seek evidence to accept or reject the hypothesis of deterministic dynamical structure on decadal time scales in the model.

Behavior of the Atmospheric Circulation Model

Before embarking on the time series analysis, a survey of qualitative behavior was made by running the model a number of times with different values of key parameters. Specifically, we varied the global mean solar forcing T_0^* and the vertical lapse rate λ. The purpose of this survey was twofold: to ascertain the robustness of chaotic behavior, and to locate any transitions from regular to chaotic dynamics which might yield some insight into the chaotic regime.

With T_0^* reduced to produce winter-like temperatures, the model was found to have quasi-periodic dynamics, with a fast period of several days and a slow period of more than a year. The fast dynamics correspond to the circulation of a low pressure region in the atmosphere which is most pronounced at higher latitudes. The slow dynamics correspond to a warmer-than-average region in the sea gradually drifting around the globe. The fast dynamics are most clearly seen in modes 4 and 5, particularly T_4 and T_5, while the slow dynamics are clear in S_1 and S_2.

As the mean solar forcing T_0^* is increased, the dynamics change first to a mild chaos with very small fluctuations of T_0. Then a dangerous bifurcation [16] occurs as a threshold value of T_0 is passed: the system makes a transient jump to a different chaotic attractor. While temperatures are still winter-like, the fluctuations in T_0 are larger, being now comparable in magnitude to the fluctuations observed at Lorenz's parameter values (which produce spring-like temperatures). Although this transition is a dangerous bifurcation, there is also evidence of a nearby transition to the spring-like chaos by type 3 Pomeau-Manneville intermittency [17, 16]. This suggests that the warmer, spring-like chaos may have attractor dimension much lower than 27. Further increasing T_0^* to the spring-like conditions of Lorenz reveals no further obvious bifurcations.

In short, parameter variations showed that chaos with substantial long-term fluctuations of T_0 is reasonably robust, and yet not far from a transition to regular, quasiperiodic dynamics with fast and slow periods.

Building an Embedding

Henceforth we consider only the reference case with $T_0^* = 273K$, as proposed by Lorenz. A previous study [14] found this chaotic attractor to be characterized by just two positive Lyapunov exponents. This could be realized by an attractor in a phase space which is locally four-dimensional. Together with the transition scenarios mentioned above, this suggests that the chaotic attractor occupies a subset of phase space of dimension much smaller than 27. Our aim is then to find a reasonably small subset of the 27 modal coefficients which faithfully represent the chaotic attractor, that is, an attractor embedding in fewer than 27 dimensions. Because the model appears to have two widely different time scales, it seems likely that such a multichannel embedding may be more efficient than the popular time-delay reconstruction from a single channel of data, using a fixed size time delay for each added coordinate. Although reconstruction of phase space from a single time series is an invaluable tool when only one time series is available, it may not be preferred when multichannel data are available. We want to see what advantage can be gained from more complete multichannel data. In the process, we shall discover an unexpected property of the most interesting modal coefficients, the global means T_0 and S_0.

Shown in Figure 5 is the result of a typical 20-year simulation in the form of time series for six of the modal coefficients, modes 0, 1, and 4 of the atmospheric temperature T and the sea surface temperature S. Mode T_4 consists predominantly of fast oscillations; the individual cycles of about two weeks duration cannot be distinguished on this time scale. Modes S_0 and S_1 show predominantly slow dynamics; note however that there is no evident correlation between S_0 and S_1. Atmospheric modes T_0 and T_1 parallel the corresponding sea surface modes, with additional fast oscillations superimposed.

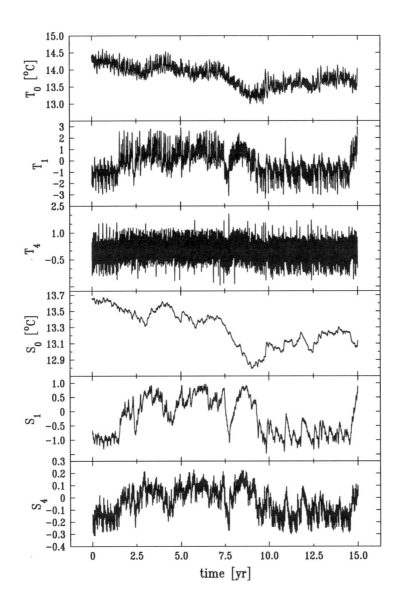

Figure 5: Selected modal coefficients of a solution of the Lorenz moist circulation model over 15 simulated years.

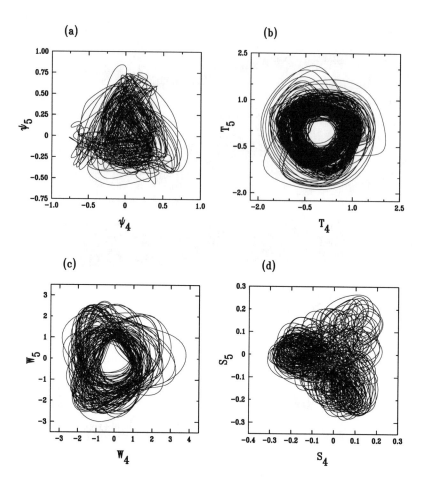

Figure 6: A solution of the Lorenz moist circulation model projected onto the planes of modes 4 and 5.

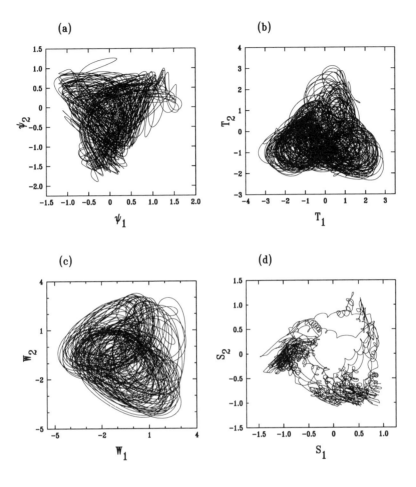

Figure 7: A solution of the Lorenz moist circulation model projected onto the planes of modes 1 and 2.

From extensive visualization of solutions in both time and space-time (using purpose-built visualization software), it is apparent that the underlying structure of the dynamics in the chaotic regime is a fast circulation most evident in modes 4 and 5, and a slower circulation most evident in modes 1 and 2. Thus any choice of modal coefficients to capture the attractor dynamics should be capable of clearly representing a two-torus, topologically the product of two circles. None of the 27 modal coefficients is by itself an angular variable; that is, any one coordinate will take the same numerical value during the ascending and the descending phase of its cycle. To distinguish the ascending and descending phases of mode 1, the complementary mode 2 must be included. Thus the representation of each circular motion (=circulation) will require a pair of coordinates.

Figure 6 shows projections of a 20-year trajectory onto the planes of modes 4 and 5 for ψ, T, W, S. One of these pairs of coordinates should be suited to describe the fast circulation. A roughly circular motion is particularly evident in T and W. Projections of the same trajectory onto the planes of modes 1 and 2 are shown in Figure 7. Here the rough circularity of the slow circulation is obscured by the superimposed fast wiggles, but it becomes more evident if one watches the trajectory being drawn very rapidly.

It is of course possible to describe a circular motion with one angular variable, for example an angle in the (T_4, T_5) plane. We have decided against this tactic for two reasons. First, we limit our analysis to picking coordinates from the 27 given, without combining in any way; while new coordinates combining the 27 might be very useful, they are beyond the scope of this study, which is our first effort at mutlichannel embedding. Second, we noted after careful consideration that an algorithm central to our analysis – the k-d tree search for nearest neighbor in phase space [18, 8] – is awkward to implement for angular variables. Considering the extra program logic required to identify $\theta = 2\pi$ with $\theta = 0$ for an angular variable in the k-d tree search, it was judged a more reliable and no less efficient strategy to require that any angular variable θ be represented by two displacements, analogous to $\sin \theta$ and $\cos \theta$.

Therefore we begin building an embedding of the attractor by considering all four mode 1,2 pairs and all four mode 4,5 pairs. We may be paying a penalty for this, using pairs of coordinates where perhaps one (angular) coordinate would suffice. According to topological embedding theorems [7, 19] any method of reconstruction in Euclidean space (including the time-delay method) must expect to pay a similar penalty.

The average local divergence rates of a 20-year trajectory using all eight complementary pairs of coefficients are shown in Figure 8(a). The pair (S_1, S_2) has the lowest divergence rates, and this would seem to be the best

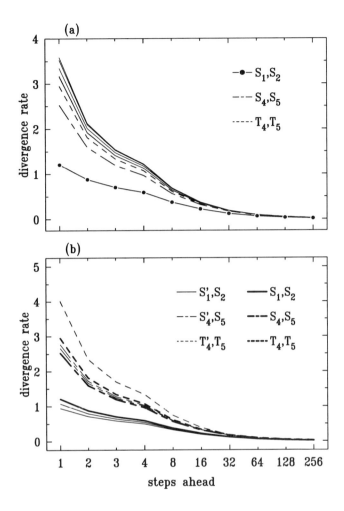

Figure 8: Average local divergence rate $\bar{s}(j)$ for trial embeddings of the Lorenz 27-variable model in two dimensions: (a) trial embeddings; (b) comparisons with surrogate channels.

choice to begin recovering the dynamical structure and near-term predictability of the attractor from data. However, the divergence rate of a subset of coordinates does not indicate the forecastability of all coordinates, but only the forecastability of the coordinates in the subset. Good forecasts for the pair (S_1, S_2) are easier because S_1 and S_2 change more slowly. One might imagine that a better diagnostic would measure divergence in 27 dimensions from nearest neighbors in two dimensions; but this strategy has problems, as we shall see below. Instead we propose the following: include a coordinate in the embedding if it yields low divergence rates, *and* yields separation rates *lower* than those with surrogate channel data substituted.

In Fig. 8(b) divergence rates for (S_1, S_2) are compared with pairs (S_1', S_2) having surrogate data substituted for S_1. The surrogate data were obtained from the same coefficient of another independent simulation using different starting conditions. Two different series of surrogate channel data were tested; one could imagine statistical tests on larger samples, but in this case the evidence is clear: both surrogates yield *lower* divergence rates than the original pair (S_1, S_2). Thus the apparent advantage of (S_1, S_2) in Fig. 8(a) is due to the persistence of S_1 and S_2, and not to the dynamical information they contain.

The second candidate from Fig. 8(a) is the pair (S_4, S_5); this pair is also compared with surrogates in Fig. 8(b). Although (S_4, S_5) diverge slightly less than with a surrogate S_4', the difference is not greater than the variation among the two surrogate samples. Referring back to the time series in Fig. 5, we see that S_4 closely parallels S_1 with some superimposed fluctuations; the favorable divergence of (S_4, S_5) is again due more to persistence than to dynamical information. We therefore pass over (S_4, S_5) and consider the third candidate in Fig. 8(a), which is (T_4, T_5). In this case, surrogates are clearly worse, as seen in Fig. 8(b). We therefore accept (T_4, T_5) as the first pair of coordinates for embedding the attractor. Referring back to Fig. 6, we see that among the mode 4,5 pairs, T has the greatest visual coherence, that is, its self-intersections are on average more oblique than those of other pairs.

Next we consider (T_4, T_5) in a four-dimensional embedding with each of the remaining pairs of modes 4,5 or modes 1,2. As seen in Figure 9(a), adding the pair (S_1, S_2) yields the lowest divergence rates. As before we check this by substituting surrogate channel data S_1', this time joined with S_2, T_4, T_5. With the fast oscillations included, the dynamical information in S_1 is now apparent. The added pair (S_1, S_2) as third and fourth coordinates is now better than the four surrogates shown in Fig. 9(b). We accept (S_1, S_2) into our embedding with (T_4, T_5).

If we had omitted the surrogate channel test, and had accepted (S_1, S_2)

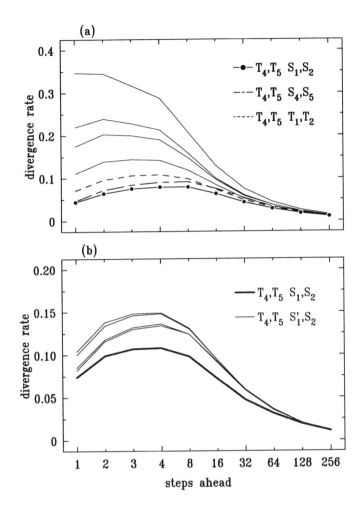

Figure 9: Average local divergence rate $\bar{s}(j)$ for trial embeddings of the Lorenz 27-variable model in four dimensions: (a) trial embeddings; (b) comparison with surrogate channels.

as the first coordinate pair, it turns out that (T_4, T_5) would have been chosen as the second pair. So in this instance the surrogate test was unnecessary. Nevertheless, we believe that the only safe strategy is to include the surrogate channel test, *even when all channels of data are known to come from a single deterministic dynamical system*. Support for this claim will appear in the next stage of coordinate selection.

Also note that (S_1, S_2) failed the surrogate test standing alone, although they passed when combined with (T_4, T_5). This differs from the results with the simple Roessler attractor, where every possible pair of coordinates passes the surrogate test even though the third coordinate is not used. Thus in at least some cases, a partial embedding or projection of a dynamical trajectory may fail to show signs of deterministic dynamical structure when diagnosed with average local divergence rates and tested against surrogate channel substitutions. It may be that this failure is a consequence of the two very different time scales in the Lorenz 27-variable model.

We next consider each of the remaining pairs of modes 1,2 or modes 4,5 with (T_4, T_5, S_1, S_2) in six-channel embeddings. Note that since we have already accepted two pairs of coordinates which should capture an underlying two-torus of fast and slow circulations, it is no longer strictly necessary to test channels in pairs. Our tactic here is to try one more pair, and then consider channels singly. As seen in Figure 10(a), the best pair to add is now (T_1, T_2); this pair passed a surrogate test (not illustrated). We then substituted coordinates singly into the six-channel set $(T_4, T_5, S_1, S_2, T_1, T_2)$. These tests were not exhaustive, but none of the single channel substitutions tested gave better divergence rates. In one case, divergence rates almost as low were obtained, by substituting the global mean sea surface temperature S_0 for S_1. However, surrogate tests in Fig. 10(b) showed that although S_1 is better that S_1', S_0 is not significantly better than S_0'.

Referring again to Fig. 5, we note that S_0 varies slowly, on a time scale like that of S_1. But S_1 passes the surrogate channel test in four- and six-channel partial embeddings. So the failure of the surrogate test for S_0 in Fig. 10(b) has a different meaning from the failure of the surrogate test for S_1 in Fig. 8(a). We interpret Fig. 10(b) as indicating that the level of the sea surface temperature S_0 is not dynamical information usable for (short-term) forecasting. Of course S_0 is generated by solving a deterministic model, so in some sense it must contain dynamical information. Indeed, when the time derivative dS_0/dt is used as a coordinate, it does pass the surrogate channel test (although its divergence rates are worse than those of S_0). In other words, these simple surrogate channel tests suggest that while changes in S_0 contain dynamical information usable for short-term forecasting, S_0 itself does not.

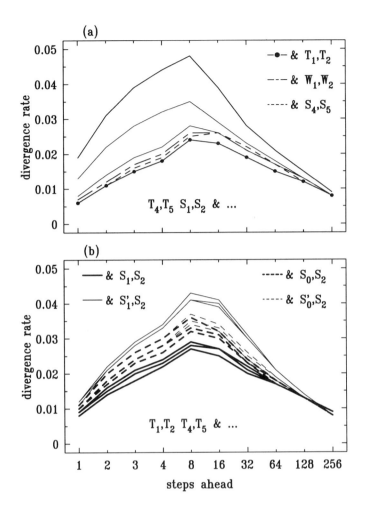

Figure 10: Average local divergence rate $\bar{s}(j)$ for trial embeddings of the Lorenz 27-variable model in six dimensions: (a) trial embeddings; (b) comparisons with surrogate channels.

This becomes understandable if one thinks of the sea in the Lorenz model as a thermal buffer, each day changing its temperature slightly by interaction with the atmosphere. Each daily change is small and deterministic, but the net level of S_0 depends on many daily changes, accumulated over times longer than the forecast horizon implied by the size of the largest Lyapunov exponent. In the process of accumulating these changes over long times, the deterministic origin of each daily change is lost, so the level of S_0 is no help in forecasting.

The apparent lack of dynamical information in S_0 has several important consequences. If one wished to reconstruct this chaotic attractor from a single time series, it would seem unwise to choose S_0. Furthermore, since T_0 parallels S_0 (with superimposed fast oscillations), T_0 also seems a poor choice; using the difference $T_0 - S_0$ rather that T_0 would probably enhance the dynamical information content for the purpose of short-term forecasting.

Moreover, even when using multichannel data as phase space coordinates to study this attractor (for example, to estimate Lyapunov exponents), it would seem that S_0 should not be used as a coordinate, and T_0 should be replaced by $T_0 - S_0$.

Finally, because the yearly average of T_0 is nearly identical with the yearly average of S_0, it becomes very difficult to imagine how the long-term fluctuations of T_0 could possibly possess any dynamical structure usable for forecasting. Nonetheless, we shall not abandon our plan of testing this hypothesis directly. But first we shall complete the multichannel embedding of the short-term dynamics.

Completing the Embedding

The best six-dimensional embedding achieves local divergence rates of about 0.025 per time step, corresponding to doubling in about 28 time steps or one nominal week. This is comparable to the time scale for the growth of perturbations, as illustrated in Figure 3. So far we have considered how to choose successive additions to the embedding coordinates; we now need a means of determining when the embedding is complete.

One popular approach is to look for saturation of an invariant: when an added coordinate no longer changes the value of an invariant quantity such as the largest Lyapunov exponent. Our crude estimates of local divergence rate are meant to approximate the largest Lyapunov exponent for sufficiently large j, so we could look for $\bar{s}(j)$ to stop changing. However this approach is not practical because extraordinarily long time series would be required to make $\bar{s}(j)$ with fixed j behave as an invariant.

In any case, a more direct determination of embedding completeness can be calculated using false nearest neighbors [8]. The idea is to determine whether points which are close neighbors in an embedding of dimension d remain close in a higher dimension, or are separated by the additional coordinate. In the latter case, the d-dimensional close neighbors only appear close due to projection to a dimension which is too low. The algorithm for this is as follows. Consider a d-dimensional embedding with coordinates $Y^d = (y_1, y_2, ...y_d)$. For each observation $Y^d(t_i)$, find the nearest neighbor in phase space $Y^d(t_{\mathcal{N}(i)})$. Then augment the embedding coordinates to $Y^{d+1} = (y_1, y_2, ...y_d, y_{d+1})$, and compare the new distance $|Y^{d+1}(t_{\mathcal{N}(i)}) - Y^{d+1}(t_i)|$ with the former distance $|Y^d(t_{\mathcal{N}(i)}) - Y^d(t_i)|$. Note that the same nearest neighbor $\mathcal{N}(i)$, determined in d dimensions, is used in both cases; only the distance formula is affected by the added coordinate. If the ratio

$$|Y^{d+1}(t_{\mathcal{N}(i)}) - Y^{d+1}(t_i)|/|Y^d(t_{\mathcal{N}(i)}) - Y^d(t_i)| \tag{9}$$

of new to old distance is large, say greater than 10, then the added coordinate has separated false nearest neighbors; otherwise the nearest neighbor is considered a true near neighbor. If the nearest neighbor is not false for any t_i, then the added coordinate is redundant.

There is also a second criterion for declaring a nearest neighbor to be false; this criterion is needed when the deterministic nature of the time series is in question. We shall discuss this second criterion below when we analyze the long-term dynamics; for the present purpose, only the criterion based on the distance ratio (9) is required. (In fact we evaluated the second criterion and confirmed that it is never satisfied in analyzing the short-term dynamics.)

The computation of false nearest neighbors has been used successfully with time delay embeddings; note that in this case coordinate $d+1$ should be the one with the *least* delay. The extension of the false nearest neighbor algorithm to multichannel embeddings is immediate. The only difference is that for a given d-dimensional embedding, every remaining channel should be tested as a possible $(d+1)$st coordinate. If each remaining channel yields no false nearest neighbors, then the embedding is complete.

Figure 11 shows the results of various false nearest neighbor computations using multichannel embeddings, and for the sake of comparison, a time delay embedding using only the T_4 time series. The numbers are expressed as percentages; for 60-year simulations the number of observations is 86400. Based on experience which shows that true nearest neighbors are occasionally identified spuriously as false, we treat fewer than 0.01% false nearest neighbors (which in this case means a count of 8 or fewer) as essentially zero.

For dimensions 1 through 5, Fig. 11 shows the effect of adding (or removing) each channel in succession from our best six-channel embedding. Thus for example the number of false nearest neighbors in the multichannel

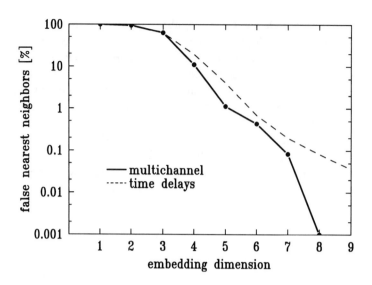

Figure 11: Percentages of false nearest neighbors for multichannel and time delay embeddings in the Lorenz model.

case for dimension 5 represents the result of adding T_2 to the five channels T_4, T_5, S_1, S_2, T_1. For dimensions 1 through 5, we have not re-evaluated alternative embeddings by comparing false nearest neighbors, although this could certainly be done. Here we accept the results of the comparisons based on local divergence; the false nearest neighbor percentages for dimensions 1 through 5 are shown for the sake of completeness. Our analytical use of false nearest neighbors is in choosing coordinates 7 and 8, corresponding to dimensions 6 and 7 in Fig. 11.

When a coordinate is added to the six-channel embedding, a number of choices yield significant false nearest neighbors. The greatest number (about 0.4%) is produced by adding W_6. If we select this as coordinate 7, and then check for false nearest neighbors produced by an eighth coordinate, only W_3 yields a significant number of false nearest neighbors (about 0.1%). From this it follows that if we now add any remaining coordinate to the eight-dimensional embedding $T_4, T_5, S_1, S_2, T_1, T_2, W_6, W_3$, only an insignificant number of false nearest neighbors will be generated. We therefore consider this eight-dimensional embedding to faithfully represent the dynamics on the attractor. According to the embedding theorems, eight is the number we may expect for an attractor which has two positive Lyapunov exponents and is therefore locally four-dimensional (and assuming the global dimension is also four).

For comparison, Fig. 11 also shows false nearest neighbors for time delay embeddings of coordinates T_4. A basic delay interval of 16 steps was chosen, which is 1/4 of the approximate period of fast oscillations. Note that even at dimension 9 the number of false nearest neighbors is still significant. This would *seem* to suggest that the attractor has more than four dimensions. However, the embedding theorems states that eight should suffice in principle for a four-dimensional attractor: it may nevertheless fail in practice. We suspect that in this instance, time delay embedding does indeed fail to embed a four-dimensional attractor in eight dimensions. The reason for this can be found in the two widely differing time scales in the model: the fast oscillation is well sampled with a delay of 16 steps, but even 10×16 does not approach the characteristic time scale of the slow oscillation.

In sum, we have used local divergence rates of nearest neighbors, comparisons with surrogate channel substitutions, and false nearest neighbors to select a subset of 8 of the 27 time series which faithfully represents the chaotic attractor, and can be used to make short-term forecasts. As a significant by-product of this analysis, we learned that one channel, the global mean sea surface temperature S_0, is not directly useful for identifying dynamical analogs and making short-term forecasts; if S_0 is to be used for this purpose, one must first differentiate it with respect to time.

Long-term Dynamical Structure in the Model

We now turn to the question of whether yearly averages of the Lorenz 27-variable model have dynamical structure on annual or decadal time scales. This question has already been discussed in the literature. Nese and Dutton [14] computed the correlation dimension from time-delay embedding using the times series T_0 and a fundamental delay of 12 years, determined from a mutual information algorithm [20,8]. The computed dimension was about 5.8. However, as noted by Ruelle [21], the correlation dimension will saturate and give an incorrect low dimension for an insufficiently long time series, even if the data are truly random. Ruelle estimates the saturation value to be $2\log_{10}(N)$; Nese and Dutton used 10,000 years of data with a delay of 12 years, so $N = 10,000/12$, and the saturation value is about — 5.8! No other evidence of dynamical structure on the long time scale was presented in [14].

McDonald [15] has analyzed annual averages of T_0 from the Lorenz model using rescaled range analysis to obtain a persistence index for various time scales. He concluded that for periods less than 20 years, the Lorenz model can be modeled as a Brownian motion with independent increments. For periods from 20 to 400 years the persistence index showed partial coherence. **McDonald concluded with appropriate caution that predictions on these time**

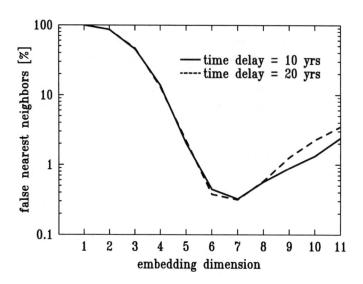

Figure 12: Percentages of false nearest neighbors for time delay embedding of the yearly averaged global mean atmospheric temperature T_0 in the Lorenz model.

scales might be possible; that is, the evidence is consistent with dynamical structure on these time scales.

False nearest neighbors can also be used to test for dynamical structure. As noted above, this requires a second criterion to identify false nearest neighbors [8]. The reason for this is that when a finite-length time series of random data is embedded in successively higher dimensions, eventually the nearest neighbor of any point will be far enough distant that it becomes virtually impossible to increase the distance by a factor of ten. Thus the second criterion for declaring a nearest neighbor false is that if its distance is larger than some preset value, we don't care what the added dimension does, the nearest neighbor is automatically counted as false. With proper normalization, this will never happen for deterministic data. We rescale each channel to the interval $[0,1]$ and take the threshold for this second criterion to be $0.02 d^{1/2}$.

Although we would like to use multichannel analysis, the annual averaging of coefficients in the Lorenz model reduces the variance of almost all coefficients to a tiny fraction of their original variance. Only the global means T_0, W_0, and S_0 retain substantial variance after taking yearly averages, and all three of these are essentially the same time series. Therefore we have little choice but to consider time delay embeddings. We have tried a delay of 10 years, as suggested by the mutual information calculation of [14], and also 20 years as indicated by the rescaled range analysis of [15]. The time series used was T_0, as in both previous studies.

The results of these false nearest neighbor calculations are shown in Fig. 12. The increase following the minimum at dimension 7 is the typical indication of data without deterministic dynamical structure. The false nearest neighbors counted for dimensions 1 through 6 are all generated by the first criterion for falseness; thereafter, most false nearest neighbors come from the second criterion. We conclude that the evidence from rescaled range analysis, while tantalizing, is not borne out by the more direct test for dynamical structure using false nearest neighbors.

A PALEOCLIMATE EXAMPLE

Finally we show by example how such methods might be applied to paleoclimate data. Here we must, at least for the present, abandon thought of any specific model which would describe dynamical structure of climate. Of course there are elaborate dynamical models of climate evolution, but when atmospheric dynamics are included, the differential equations model the evolution in steps of hours or days, and nearby trajectories diverge in weeks or at most months, as in the low-order Lorenz model. Here we wish to pose a different question, analgous to the hypothesis of long-term dynamical structure in the Lorenz model: Is the evolution of climate governed by dynamical laws, known or unknown, which determine the state one or a few years forward given the present state, expressed in terms of annual or seasonal means.

We emphasize that we know of no persuasive argument that climate has such a dynamical structure (on yearly time scales) at all. In fact, the rate of divergence of weather conditions seems (at least superficially) to make this dynamical structure of climate an unlikely hypothesis. Nevertheless, it is not ruled out. If true, or even partly true, it would be of great consequence: as just one example, it might permit earlier detection of a greenhouse signal manifested as a departure in (or from) dynamical structure. In any case, we are simply posing this as a hypothesis, to be tested using multichannel time series analysis.

The growing body of paleoclimate proxy data with annual resolution makes it possible to consider whether year-to-year climatic fluctuations can be described by a deterministic dynamical rule, in the sense stated above, by examining observed data. Furthermore, the absolute dating of annual resolution data means that the concurrence of two or more independent time series can be established. This concurrence is an essential prerequisite for asking whether such multichannel data represent phase space coordinates for a possible deterministic dynamical rule.

For our example, we shall use two paleoclimate proxy data series of very different origin. For a more thorough study, additional data series should of course be included. One series used here is the annual oxygen isotope ratio in cores from the Quelccaya ice cap in Peru published by Thompson and Mosley-Thompson [22], extending from 1476 to 1984. The other series is a tree ring index developed by Briffa et al. [23] from Fennoscandian trees, extending from A.D. 500 to 1980, and kindly furnished to the author by Prof. Briffa. These two data series were treated as possible phase space coordinates for dynamical structure, if it exists. The notion underlying this choice is that these two proxies would represent different and complementary modes or degrees of freedom of a hypothesized dynamical structure. Of course any climatological insight which would bear on the suitability of these or other data series as reflecting the state of the climate should be considered; see for example [24]. For present illustrational purposes, we take these two series to represent the state of the art.

These two data series are analyzed on two slightly different time scales. First, we test for dynamical structure on the yearly time scale. Both data series exhibit large fluctuations from year to year; for this reason, a smooth evolution on yearly time scales as with differential equations is not an appropriate hypothesis, and instead we must suppose a system in which time passes in discrete units as in the iterated function (1). Second, we test for dynamical structure on multiyear time scales, taking running means to smooth the data; this gives somewhat less ragged trajectories which might correspond to a differential equation model. The numerical evaluation of local divergence rates is essentially the same in both cases, with one small difference: with the smoothed data we determine nearest neighbors by interpolating between successive data points, while with the raw yearly data, there is no interpolation.

In the previous examples of synthetic data from simulations, there was little harm in generating surrogates by cutting a long trajectory into two disjoint segments: small sample size was not a problem, as the simulation can always be extended. However, with limited and precious paleoclimate data, cutting the time series in half would seem extravagant: one would prefer to test the full length of available data against surrogates of equal length.

Another means of generating surrogate channels is by randomization of the given data. Substituting data from a random number generator is too crude: we want our surrogates to lack dynamical structure, but be able to pass superficially for the real data. One algorithm for achieving this is the following: take the complex Fourier transform of the data, then randomize

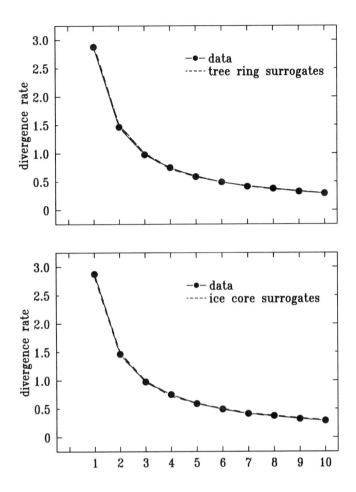

Figure 13: Average local divergence rate $\bar{s}(j)$ of a paleoclimate proxy data trajectory from 1476-1980. Phase space coordinates are oxygen isotope ratio from Quelccaya ice core, and Fennoscandian tree ring index. Above, $\bar{s}(j)$ vs. j for the paleoclimate trajectory and three trajectories with surrogate tree ring data; below, the paleoclimate trajectory and three trajectories with surrogate ice core data.

the phases of the complex coefficients $a(\omega)$ in the frequency domain. In order that the inverse transform yield a real time series, the phases must be randomized under the constraint that $a(\omega) = a^*(-\omega)$. The inverse transform of data randomized in this manner will be a surrogate having the same power spectrum as the original data, but with any dynamical structure removed.

This is the procedure we adopt for generating surrogates for both the ice core and the tree ring data. It is not the only, or necessarily the best procedure; see [10, 11, 25] for further discussion.

Figure 13 shows local divergence rates $\bar{s}(j)$ for a two-dimensional trajectory whose coordinates are the Quelccaya oxygen isotope ratio and the Fennoscandian tree ring index; the values of $\bar{s}(j)$ are indicated by dots and connected with solid lines. Also shown are the divergence rates for three different surrogate substitutes for each coordinate, plotted as broken lines. In the upper plot, surrogates are substituted for the tree ring data. The values of $\bar{s}(j)$ for surrogates fall very slightly above the values for the data at $j = 1$ and $j = 2$; we can easily imagine that with more than three surrogate substitutions, their range would include the $\bar{s}(j)$ of the original data. For larger j, the $\bar{s}(j)$ for the data lie within the range for the three surrogates. In the lower plot, surrogates are substituted for the ice core data; all the $\bar{s}(j)$ for the original data are within the ranges for the surrogates. We conclude that these tests show no evidence of dynamical structure in the data on yearly time scales.

The same two time series were then smoothed and re-tested for local divergence on multiyear time scales. Seven-year running means were used, based on the presence of a strong 14-year component in spectrum of the ice core data. We also used eleven- and thirteen-year running means. When generating surrogates for smoothed data, we randomize phases of the raw data and then smooth, rather than randomizing smoothed data; this is in accord with the recommendation of [26] to randomize before filtering. Results for the running means were similar to the unsmoothed results. The divergence rates for the data again fall within or are very close to the range for the surrogates. We conclude that there is also no evidence for dynamic structure in the data on multiyear time scales.

The development and application of dynamical systems approaches to time series analysis is still in its infancy. In due course, with deeper insight into climate and perhaps some serendipity, these multichannel methods should achieve successes with climate data comparable to the recent success analyzing single time series [3].

Acknowledgement. The author is indebted to Henry Abarbanel, Keith Briffa, Caroline Isaacs and Ed Lorenz for stimulating discussions. This work was supported by the CHAMMP initiative of the Office of Health and Environmental Research, U.S. Department of Energy under Contract No. DE-AC02-76CH00016.

References

[1] Abraham, R.H., and C.D. Shaw, *Dynamics: the Geometry of Behavior*. Redwood City: Addison-Wesley, 1992.

[2] Thompson, J.M.T. and Stewart, H.B., *Nonlinear Dynamics and Chaos*. Chichester: John Wiley and Sons, 1986.

[3] Abarbanel, H.D.I., and Lall, U., "Nonlinear dynamics of the Great Salt Lake: system identification and prediction," unpublished manuscript (1994).

[4] Abarbanel, H.D.I., *Analysis of Observed Chaotic Data*. New York: Springer-Verlag, to be published.

[5] Lorenz, E.N., "Atmospheric predictability as revealed by naturally occurring analogues," *Journal of the Atmospheric Sciences* **26**, 636-646 (1969).

[6] Packard, N.H., Crutchfield, J.P., Farmer, J.D., and Shaw, R.S., "Geometry from a time series," *Physical Review Letters* **45**, 712-716 (1980).

[7] Takens, F., "Detecting strange attractors in turbulence," in *Dynamical Systems and Turbulence*, D.A. Rand and L.-S. Young, editors, Lecture Notes in Mathematics 898, New York: Springer-Verlag, 1980, pp. 366-381.

[8] Abarbanel, H.D.I., Brown, R., Sidorowich, J.J., and Tsimring, L.Sh.T., "The analysis of observed chaotic data in physical systems," *Reviews of Modern Physics* **65**, 1331-1392 (1993).

[9] Roessler, O.E., "An equation for continuous chaos," *Physics Letters* A **57**, 397-398, (1976).

[10] Theiler, J., Eubank, S., Longtin, A., Galdrikian, B., and Farmer, J.D., "Testing for nonlinearity in time series: the method of surrogate data," *Physica* D **58**, 77-94 (1992).

[11] Takens, F., "Detecting nonlinearities in stationary time series," *International Journal of Bifurcation and Chaos* **3**, 241-256 (1993).

[12] Lorenz, E.N., "Formulation of a low-order model of a moist general circulation," *Journal of the Atmospheric Sciences* **41**, 1933-1945 (1984).

[13] Stewart, H.B., "The Lorenz low-order model of moist atmospheric circulation I: Regimes and Birfurcations," unpublished manuscript (1995).

[14] Nese, J.N., and Dutton, J.A., "Quantifying predictability variations in a low-order ocean-atmosphere model: a dynamical systems approach," *Journal of Climate* **6**, 185-204 (1993).

[15] McDonald, G., "Persistence in climate," Report JSR-91-340, unpublished (1992).

[16] Thompson, J.M.T., Stewart, H.B., and Ueda, Y., "Safe, explosive, and dangerous bifurcations in dissipative dynamical systems," *Physical Review* E **49**, 1019-1027 (1994).

[17] Pomeau, Y., and Manneville, P., "Intermittent transition to turbulence in dissipative dynamical systems," *Communications in Mathematical Physics* **74**, 189-197 (1980).

[18] Farmer, J.D., and Sidorowich, J.J., "Exploiting chaos to predict the future and reduce noise," in *Evolution, Learning, and Cognition*, Y.C. Lee, editor, Singapore: World Scientific, 1988, pp. 277-330.

[19] Sauer, T., Yorke, J.A., and Casdagli, M., "Embedology," *Journal of Statistical Physics* **65**, 579-616 (1991).

[20] Fraser, A.M., and Swinney, H.L., "Independent coordinates for strange attractors from mutual information," *Physical Review* A **33**, 1134-1140 (1986).

[21] Ruelle, D., "Deterministic chaos: the science and the fiction," *Proceedings of the Royal Society of London* A **427**, 241-248 (1990).

[22] Thompson, L.G., and Mosley-Thompson, E., "One-half millenia of tropical climate variability as recorded in the stratigraphy of the Quelccaya ice cap, Peru," in *Aspects of Climate Variability in the Pacific and the Western Americas*, D.H. Peterson, editor, Geophysical Monograph 55, Washington, D.C.: American Geophysical Union, 1989, pp. 15-31.

[23] Briffa, K.R., Jones, P.D., Bartholin, T.S., Eckstein, D., Schweingruber, F.H., Karlen, W., and Zetterberg, P., "Fennoscandian summers from AD 500: temperature changes on short and long timescales," *Climate Dynamics* **7**, 111-119 (1992).

[24] Cole, J.E., Rind, D., and Fairbanks, R.G., "Isotopic responses to interannual climate variability simulated by an atmospheric general circulation model," *Quaternary Science Reviews* **12**, 387-406 (1993).

[25] Kennel, M.B., and Isabelle, S., "Method to distinguish possible chaos from colored noise and to determine embedding parameters," *Physical Review* A **46**, 3111-3118 (1992).

[26] Prichard, D., "The correlation dimension of differenced data," *Physics Letters* A **191**, 245-250 (1994).

PICTURE NOT AVAILABLE

H. Bruce Stewart received his B.S. degree with distinction from Stanford and his Ph.D. degree in mathematics from the University of California at Berkeley. He has been a mathematician at Brookhaven National Laboratory for the past fifteen years. His research interests include nonlinear oscillators, low-dimensional chaos, generic bifurcations, and multifield numerical models of interpenetrating multiphase fluid flow. He is the co-author with J.M.T. Thompson of *Nonlinear Dynamics and Chaos*, John Wiley & Sons, 1986, and the animator of *The Lorenz System*, a 16mm computer-generated film describing a famed example of chaotic dynamics.

Coherent Structures Amidst Chaos: Solitons, Fronts, and Vortices

David K. Campbell

Department of Physics, University of Illinois, Urbana-Champaign
1110 W. Green Street, Urbana, Illinois 61801

Abstract. I introduce the concept of "coherent structures"– localized, persistent, propagating nonlinear waves – and argue that they are ubiquitous in spatially extended nonlinear systems. I discuss various specific forms of coherent structures – solitons, wave fronts, vortices — and illustrate how they arise in physics, chemistry, biology, and physiology.

INTRODUCTION

Most of the articles in the present volume deal with "deterministic chaos," the incredibly complex, irregular behavior in time exhibited by even simple nonlinear systems. This emphasis is hardly surprising, for both the scientific and lay communities have been first stunned and then fascinated by the strikingly counter-intuitive result that a system governed by deterministic laws can exhibit effectively random behavior. Perhaps less well known, but equally counter-intuitive, is another remarkable characteristic of nonlinear systems: the emergence of highly ordered, persistent, nonlinear waves – "coherent structures" – in seemingly very complicated, spatially extended nonlinear systems. In a sense, these coherent structures represent the "flip side" of nonlinearity, their surprising regularity countering the unexpected irregularity of chaos. In this short introduction to the topic, I begin with a brief sampler of coherent structures – some quite familiar, some less so – from the natural world. I then discuss "solitons," which are the paragons of coherent structures. The discovery of solitons has been characterized as one of the most significant developments in mathematical physics in the last half century.(1) The implications for certain future technologies may be equally profound, as I will illustrate with an example from telecommunications. After presenting a short interlude making more precise the idea that chaos and solitons are the yin-yang of nonlinearity, I introduce the more general concept of "coherent structures." Finally, I examine some particular forms taken by coherent structures in chemistry, biology, and physiology, touching on several examples that should be of particular interest to the medical community. This is far too ambitious a set of topics to be presented in detail in the limited space available, but I hope to be able to pique the interest of many readers and to guide them to fuller discussions in the literature.(2)

A SAMPLER OF COHERENT STRUCTURES

From the Red Spot of Jupiter through clumps of electromagnetic radiation in turbulent plasmas to pulses of light in optical fibers the size of a human hair, spatially localized, long-lived, wave-like excitations abound in nonlinear systems. These localized excitations reflect a remarkable and surprising orderliness in the midst of complex behavior – the "coherent structures amidst chaos" of our title – and their ubiquitous role in both natural nonlinear phenomena and mathematical models has led to the recognition of coherent structures as one of the central paradigms of nonlinear science. Coherent structures represent the natural modes for understanding the time-evolution of many nonlinear systems and often dominate the long-time behavior of these systems.

To illustrate this, let me begin by recalling one of the most familiar (and beautiful !) examples in nature: the giant Red Spot of Jupiter. This feature, first observed from earth in the late 17th century, is a swirling vortex that has remained remarkably stable in the turbulent cauldron of Jupiter's atmosphere. It represents a coherent structure on a scale of about 4×10^8 meters, or roughly the distance from the earth to the moon.

At the terrestrial level, certain classes of ocean waves form coherent structures that propagate essentially unchanged for thousands of miles. The celebrated tsunamis (harbor waves) that occasionally ravish the coast of Japan are one such example, and other less dramatic nonlinear ocean waves have been observed in the Andaman sea near northern Sumatra(3) and in near-shore waves off the coast of Oregon.(1)

In the laboratory, nonlinear coherent structures are readily made and observed in water wave tanks.(4) On still smaller scales, localized coherent structures are the natural modes of propagation of intense light pulses in optical fibers(5) and, at the atomic level, the ground states of certain compounds are "charge density waves," which can be viewed as "frozen" patterns of nonlinear coherent structures on a scale of 10^{-9} meters.(6)

SOLITONS: THE PARAGON OF COHERENT STRUCTURES

In the previous section we identified nonlinear coherent structures in nature on scales ranging from 10^8 to 10^{-9} meters—seventeen orders of magnitude! Clearly these excitations are an essential aspect of nonlinear phenomena. It is therefore very gratifying that the past twenty years have seen a veritable revolution(1) in the understanding of coherent structures. The crucial event that brought on this revolution was the discovery (and naming), by Norman Zabusky and Martin Kruskal(7) in 1965, of the remarkable "soliton." In a sense, solitons represent the purest form of the coherent structure paradigm and thus are a natural place to begin our technical discussion.

To define a "soliton" precisely, consider the motion of a wave described by an (in general nonlinear) equation. A "traveling wave" solution to such an equation is one that depends on the space (x) and time (t) variables only through the combination $\xi = x - vt$, where v is the constant velocity of the wave. The traveling wave moves through space without changing its shape and in particular without breaking apart (technically, without "dispersing"). If the traveling wave is a localized single pulse, it is called a "solitary wave." A soliton is a solitary wave with the crucial (and *very* unexpected) additional property that it preserves its form *exactly* even when it interacts with other nonlinear waves.

Motivated by some puzzling results obtained by Enrico Fermi, Stan Ulam, and John Pasta in the early days of digital computers, (2) Zabusky and Kruskal studied the nonlinear partial differential equation

$$\frac{\partial u}{\partial t} + u\frac{\partial u}{\partial x} + \frac{\partial^3 u}{\partial x^3} = 0. \tag{1}$$

Appropriately enough, this "Korteweg-deVries" – for short, KdV – equation had first been derived in 1895 (8) as an approximate description of water waves moving in a shallow, narrow channel. Indeed, in the direction of their motion, the nonlinear waves in the Andaman sea mentioned above are described quite accurately by this equation.

At the risk of introducing a bit of mathematics unfamiliar to some, let me discuss how one can find, completely *analytically*, the coherent structures in this equation. We know that we must seek a localized solution $u_s(\xi)$ that depends only on $\xi = x - vt$. If we make this substitution in the above equation, it becomes an ordinary differential equation in ξ. This equation can be solved explicitly by elementary integration, and, for solutions that vanish at infinity, one finds

$$u_s(x,t) = 3v\text{sech}^2\frac{\sqrt{v}}{2}(x - vt). \tag{2}$$

where $\text{sech}(x) \equiv 2/(e^x + e^{-x})$, so that the solution vanishes exponentially fast away from its center at $\xi \equiv x - vt = 0$. Serious readers can verify, by direct differentiation, that u_s solves the KdV equation. This solution clearly describes a solitary wave moving with constant velocity v. Moreover, the amplitude of the wave is proportional to v, and its width is inversely proportional to $\sqrt{v}$. The faster the wave goes, the narrower it gets. This relation between the shape and velocity of the wave reflects the nonlinearity of the KdV equation.

Intuitively, we can understand the existence of this solitary wave as resulting from a delicate balance in the KdV equation between the (linear) dispersive term

$$\frac{\partial^3 u}{\partial x^3}$$

which tends to cause an initially localized pulse to change shape and break up (disperse) as it moves, and the nonlinear term

$$u\frac{\partial u}{\partial x} \equiv \frac{1}{2}\frac{\partial(u^2)}{\partial x}$$

which tends to increase the pulse where it is already large and hence to bunch up the disturbance.

Although the solution u_s is by inspection a coherent structure, is it a soliton? In other words, does it preserve its form when it collides with another solitary wave? Since the analytic methods of the 1960s could not answer this question, Zabusky and Kruskal adopted an "experimental mathematics" approach and performed computer simulations of the collision of two solitary waves with different velocities. Their expectation was that the nonlinear nature of the interaction would break up the waves, causing them to change their properties dramatically and perhaps to disappear entirely. When the computer gave the startling result that the coherent structures emerged from the interaction unaffected in shape, amplitude, and velocity, Zabusky and Kruskal coined the term "soliton," a name reflecting the particle-like attributes of this nonlinear wave and patterned after the names physicists traditionally give to atomic and subatomic particles.

In the years since 1965, research has revealed the existence of solitons in a host of other nonlinear equations, primarily but not exclusively in one spatial dimension. Significantly, the insights gained from the early experimental mathematical studies have had profound impact on many areas of more conventional mathematics, including infinite-dimensional analysis, algebraic geometry, partial differential equations, and dynamical systems theory. Interested readers can pursue the more mathematical aspects of these fascinating objects in any of several excellent monographs. (9, 10, 11)

From many perspectives, nonlinear partial differential equations containing solitons are quite special. Nonetheless, there is a surprising mathematical diversity to these equations. Further, it is often the case (as we shall indicate below) that the *same* soliton equation arises in quite different physical contexts. The combination of the diversity and the multi-disciplinary applicability of solitons is reflected in the corresponding variety of real-world applications to problems in the natural sciences and engineering. In fiber optics, Josephson transmission lines, nerve fibers, conducting polymers and other quasi-one-dimensional systems, and plasma cavitons– as well as the surface waves in the Andaman sea ! – the prevailing mathematical models are slight modifications of soliton equations. There now exist several numerical and analytic perturbation techniques for studying these "nearly" soliton equations, and one can use these to describe quite accurately the behavior of real physical systems.

One specific, decidedly practical illustration of the application of solitons concerns long-distance communication by means of light pulses in optical fibers. At low intensities, light pulses in optical fibers propagate linearly but dispersively. This dispersion tends to degrade the signal, and, as a consequence, expensive repeaters must be added to the fiber at regular intervals to reshape and thus reconstruct the pulse. If the intensity of the light transmitted through the fiber is substantially increased, the propagation becomes nonlinear and coherent structures (solitary wave pulses) are formed. In fact, these solitary waves are very well described by the solitons of the "nonlinear Schrödinger" (NLS) equation, another of the celebrated soliton-bearing nonlinear partial differential equations. In terms of the (complex) electric field amplitude describing the light pulse, $E(x,t)$, this equation has the form

$$i\frac{\partial E}{\partial t} + \frac{\partial^2 E}{\partial x^2} + |E|^2 E = 0. \tag{3}$$

Before discussing the soliton solution to the NLS equation, as an illustration of the occurrence of the same nonlinear equation in different physical contexts, let me mention that in addition to arising in optical fibers, the NLS (or closely related variants of it) arise in the description of (1) "cavitons" in turbulent plasmas; (2) "polarons" in solid state systems; (3) "Davydov solitons" in biological systems; and (4) self-consistent (Hartree-Fock) theories of atomic and nuclear structure. This remarkable range of applications correctly reflects the importance of the NLS equation.

The solution corresponding to the nonlinear soliton pulse in the optical fiber is given by

$$E_s(x,t) = \left(\frac{2\omega + v^2}{2}\right) e^{i\omega t + vx/2} \text{sech}\left(\left(\frac{\omega + v^2}{4}\right)^{1/2}(x - vt)\right) \tag{4}$$

Although it is somewhat more challenging than in the case of the KdV equation, the dedicated reader can verify, again by direct differentiation, that E_s solves the NLS equation exactly.

In the idealized limit of no dissipative energy loss, the solitons propagate without degradation of shape; these localized coherent structures are indeed the natural, stable modes for the propagation of high-intensity light in the fiber. By controlling the timing of the laser that initiates the pulses, one can send a coded sequence of solitons – the absence of soliton in a given time window being a "0", the presence a "1" – and thus create any message with a truly digital signal; this is the essence of soliton telecommunications, which may revolutionize that industry in the next century.(5),(12)

Notice that again one sees the intrinsically nonlinear characteristic of the solitons in the relation between the amplitude (and hence the energy) and the width. In real fibers, where dissipative mechanisms cause solitons to lose energy, the individual soliton pulses therefore broaden (but do not disperse). To restore the lost energy and maintain the separation between solitons necessary for the integrity of the signal, one needs to add (cheap) amplifiers (not expensive repeaters !). Exciting recent technological developments have made it possible to use an all-optical amplifier based on the properties of erbium-doped optical fibers pumped by semiconductor lasers. Laboratory experiments have now demonstrated(12) that this approach is indeed viable, and that an all-optical system with amplifier spacings of 50 kilometers can use solitons of 20-40 picoseconds duration to send information for thousands of kilometers at a bit rate of over 20 gigabits/sec and with negligible error rates.(12) Although advances in competing linear technologies have thus far kept the soliton-based systems in the developmental laboratories, the prospects for their deployment in the field in the future are excellent.

CHAOS AND INTEGRABILITY: THE YIN AND YANG OF NONLINEARITY

The obvious qualitative contrast between the irregularity of chaos and the regularity of solitons can be made precise by mathematical formulation of dynamical systems theory. Let me thus make a brief excursion into this important and wide-reaching subject.

In essence, a dynamical system is a set of objects – physical or mathematical – that change in time according to a well-defined rule. Loosely speaking, the number of objects determines the number of "degrees of freedom" of the system. An example quite familiar from elementary physics is the simple plane pendulum, for which Newton's second law gives the explicit nonlinear ordinary differential equation (ODE) describing the time evolution in terms of the angle, θ, that the pendulum makes with the vertical:

$$\frac{d^2\theta}{dt^2} = -\frac{g}{\ell}\sin\theta, \tag{5}$$

where g is the acceleration due to gravity and ℓ is the length of the pendulum. This is a "one-degree-of-freedom" dynamical system, and can be solved completely and explicitly by analytic means: in technical terms, it is "completely integrable." All of the possible motions of the pendulum – lying at rest, oscillating back and forth (libration), or twirling around (rotation) – are very regular and can be described in simple analytic terms. If, however, we add to the pendulum some damping – as could be caused by mechanical friction – and a periodic driving force –as one might provide to prevent friction from stopping the pendulum – then the nonlinear equation of evolution can be written

$$\frac{d^2\theta}{dt^2} + \alpha\frac{d\theta}{dt} + \frac{g}{\ell}\sin\theta = \Gamma\cos(\Omega t), \tag{6}$$

where α reflects the strength of the damping and Γ the strength of the driving with frequency Ω. This simple change renders the system no longer completely integrable, and indeed for certain values of the parameters – $\alpha, \Gamma, \Omega, g$ and ℓ – and certain initial conditions, the simple pendulum can exhibit chaos, with all the attendant complexity and irregularity of motion described in many of the other contributions to these proceedings.

Most dynamical systems discussed in elementary physics texts involve a few degrees of freedom – conceptually, they can be thought of as a few pendula, somehow coupled together – and the generalizations of the equations of motion above involve several coupled ODEs. But how does one count the number of degrees of freedom in the spatially extended nonlinear systems that we are discussing? The best intuitive picture is to think of a separate nonlinear oscillator – such as a pendulum – at *every* point along a line (one dimensional space), in a plane (two dimensional space) or in a volume (three dimensional space). In each case, the "number" of such "points" (and hence oscillators) is *infinite*, so that, in dynamical systems terminology, spatially extended nonlinear systems are "infinite degree of freedom" systems, and the corresponding dynamical equations are *partial* differential equations (PDEs) – such as the KdV and NLS equations shown above – rather than ODEs. One would naturally expect that an infinite degree of freedom system would have to behave in a more complicated manner than a finite degree of freedom system. The "miracle of the soliton"(1) is that this expectation is false: for soliton systems the corresponding PDEs are in fact *completely integrable, infinite degree of freedom systems* and the resulting motions are everywhere regular, with no possibility of chaos. An elegant mathematical understanding of this extraordinary behavior exists(9),(11),(10), but to describe it here would take us too far afield. At its essence is the fact that the complete integrability of the soliton equations implies the existence of an infinite number of conserved quantities, and the invariance of solitons under interactions can be understood as a consequence of these conservation laws.

In fact, most dynamical systems are neither completely integrable nor completely chaotic but have possible motions that include both regular and irregular regions. Thus in a generic spatially extended nonlinear system, any localized solitary waves are likely not to be solitons, but rather more general coherent structures. Vortices in fluids, chemical-reaction waves and nonlinear diffusion fronts, shock waves, dislocations in metals, and bubbles and droplets are all instances of coherent structures. As in the case of the solitons, the existence of these structures results from a delicate balance of nonlinear and dispersive forces. In contrast to solitons, however, these more general coherent structures typically interact strongly and do not necessarily maintain their form or even their separate identities for all times. Fluid vortices may merge to form a single coherent structure, equivalent to a single larger vortex. Interactions among shock waves lead to diffraction patterns of incident, reflected, and transmitted shocks. Droplets and bubbles can interact through merging or splitting. Despite these non-trivial interactions, the coherent structures can be the nonlinear modes in which the dynamics is naturally described, and they may dominate the long-time behavior of the system. To exemplify more concretely the essential role of these general coherent structures in nonlinear systems, let me focus on two broad classes of such structures: vortices and fronts.

The importance of vortices in complicated fluid flows and turbulence has been appreciated since ancient times. The giant Red Spot, mentioned above, is a well-known example of a fluid vortex, as are tornados in the earth's atmosphere, large ocean circulation patterns called "modons" in the Gulf Stream current, and "rotons" in liquid helium. In terms of practical applications, the vortex pattern formed by a moving airfoil is immensely important. Not only does this pattern of vortices affect the fuel efficiency and performance of the aircraft, but it also governs the allowed spacing between planes at take-off and landing: strong vortices generated by the large planes can literally hurl smaller aircraft into the ground. More generally, vortices are the coherent structures that make up the turbulent boundary layer on the surfaces of wings or other objects moving through fluids. Further, methods based on idealized point vortices provide an important approach to the numerical simulation of certain fluid flows.

The existence of fronts as coherent structures provides yet another illustration of the essential role of nonlinearity in the physical world. Linear diffusion equations cannot support wave-like solutions. In the presence of nonlinearity, however, diffusion equations can have traveling wave solutions, with the propagating wave front representing a transition from one state of the system to another. Thus, for example, in chemical reaction-diffusion systems – which we will examine in more detail in the following section – one can find traveling wave fronts separating reacted and unreacted species. Often, as in flame fronts or in internal combustion engines, these traveling chemical waves are coupled with fluid modes as well. Concentration fronts arise in the leaching of minerals from ore beds. Moving fronts between infected and non-infected individuals can be identified in the epidemiology of diseases such as rabies. In tertiary oil recovery processes, (unstable) fronts between the injected water and the oil trapped in the reservoir control the effectiveness of the recovery process. In sum, coherent structures are ubiquitous and important in the physical world. In the next section, we will explore in some detail a few important examples from chemistry, biology, and physiology.

COHERENT STRUCTURES IN CHEMISTRY, BIOLOGY, AND PHYSIOLOGY

To understand the coherent structures that arise in spatially extended chemical, biological, and physiological systems, we must begin with some of the simpler nonlinear dynamical phenomena that occur in these systems. Consider a chemical reaction in a *well-stirred* vessel; although the vessel is clearly spatially extended, because of the stirring, any localized coherent structures will be quickly destroyed by the mixing, so that the system can be considered spatially homogeneous. Once one assumes that the concentrations of chemicals are uniform in space, then the chemical system is described by a set of coupled ODEs for describing various possible reactions.

Usually one thinks of chemical interactions proceeding steadily and smoothly toward equilibrium. But our dynamical systems experience with coupled nonlinear ODEs suggests that oscillations and even chaos might be found. In fact, the possibility of oscillating chemical reactions was suggested theoretically in the early days of nonlinear dynamics(13) and was first reported in experiment not long after.(14) But the idea of intrinsic chemical oscillations was *very* controversial and was extensively criticized; the experimental observations were attributed to "dirt effects" (impurities) or external perturbations (photosensitive reactions affected by stray light). Chemical oscillations were even declared to be impossible, a violation of the laws of thermodynamics. One of the strongest statements of this sort was made in(15), in which it was asserted

> "However, if the individual processes are microscopically reversible (which is necessary if a final equilibrium state is to be attained),

a transformation can always be found which symmetries (sic) the Master equation for an isolated system, so that the eigenvalues are necessarily real... Consequently, a periodic reaction cannot occur in the homogeneous phase in a thermodynamically closed system."

The hidden assumption in this "proof" of the impossibility of chemical oscillations is effectively the *linearization* around equilibrium; that is, the master equation approach employed effectively assumed that the system was sufficiently close to equilibrium that only linear deviations need be considered. But this is obviously not necessarily true, if one starts with concentrations very far from the equilibrium values. In particular, even if the concentrations of the initial reactants decrease continuously and those of the final products increase continuously, the concentrations of *intermediates* (which concentrations will go to zero in the final equilibrium) can undergo oscillations.(16) There are now many known examples of oscillating chemical reactions, and a systematic approach to understanding and predicting these oscillating systems exists.(17) Further, demonstrations of chemical oscillations are a standard experiment in elementary chemistry courses.(16)

To give an important specific example, perhaps the most celebrated chemical oscillator is the Belusov-Zhabotinsky (BZ) reaction, which involves the oxidation of an organic material (typically malonic acid) by bromate ions in a strongly acidic aqueous solution and in the presence of a metallic catalyst. Including the transient intermediate species, the actual chemical mechanism is quite complicated, but there is a very simplified model (known as the "Oregonator"(16)) which captures the essence of the observed oscillations. In terms of the concentrations of $HBrO_2 (\equiv c_1)$, $Br^- (\equiv c_2)$, and the metal catalyst (either cerium $2Ce(IV)$ or ferroin, and called c_3), the equations are

$$\frac{dc_1}{dt} = \alpha c_2 - \beta c_1 c_2 + \gamma c_1 - \delta c_1^2$$

$$\frac{dc_2}{dt} = -\alpha c_2 - \beta c_1 c_2 + \nu c_3 \qquad (7)$$

$$\frac{dc_3}{dt} = \gamma c_2 - \epsilon c_3$$

where the Greek letters represent reaction rate constants (or combinations of reaction rate constants with concentrations of other species that can be considered constant on the time scale of single oscillations). Although I do not have the space to analyze these equations in detail, I feel that exhibiting their explicit form is nonetheless useful, for it demonstrates just how simple the dynamical equations (the nonlinearities are just the $c_1 c_2$ and c_1^2 terms) for these chemical oscillations are. It is unfortunate that the static pages of this volume cannot illustrate the striking visual image of a sequence of bulk oscillations, which in the case of the ferroin catalyst are between red and blue, that occurs in the BZ reaction in a large container. It provides an impressive and lasting reminder of the dangers of relying on linear intuition !

As our previous discussions suggest, to observe coherent structures in chemical and biological systems, we must have inhomogeneity (*i.e.*, non-uniform concentrations) and the chemical reactants must be able to move through the system. In chemical and biological systems, diffusion is the most common means of transport, so the natural equations to study are *nonlinear* reaction-diffusion equations. A classic example of these equations was formulated in 1936 by Fisher(18) in the context of the propagation of a mutant gene and has since also been used (among other applications) to model the motion of both chemical reaction fronts and the spatial spread of infectious diseases. In its simplest form, Fisher's equation is

$$\frac{\partial u}{\partial t} = \frac{\partial^2 u}{\partial x^2} + u(1-u). \tag{8}$$

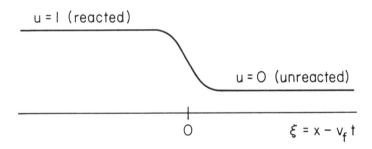

Figure 1 A moving front solution to Fisher's equation

This equation admits a coherent structure (solitary wave) solution which, as the spatial variable x runs from $-\infty$ to ∞, interpolates between the (stable) constant state $u = 1$ (which represents the reacted species or infected individuals) and the (unstable) constant state $u = 0$ (which in the chemical context represents the unreacted species and in the epidemiological context represents uninfected individuals). Although one can prove the existence of stable "front" solutions for a range of velocities, an instructive explicit form of this moving front solution, $u_f(x,t)$, for a particular velocity was first discovered in 1979(19) (see also (20)) and is given by

$$u_f(x,t) = (1 + exp(a_f(x - v_f t))^{-2}, \tag{9}$$

with $a_f = 1/\sqrt{6}$ and $v_f = 5/\sqrt{6}$. The form of this front solution is sketched in Fig. 1.

More general reaction-diffusion equations allow for reactions among multiple species, such as concentrations of different chemicals. In a monumental work in the early 1950's, Alan Turing proposed(21) that reaction-diffusion equations could form a chemical basis for "morphogenesis," the creation of inhomogeneous structures and patterns in an originally homogeneous system. Although this brilliant insight has guided much of mathematical biology during the intervening decades, technical complications prevented the experimental observation of the explicit mechanism and patterns suggested by Turing until very recently. Nonetheless, it is now generally agreed that these "Turing patterns" have been observed in both chemical(22),(23) and biological(24) systems.

Rather than discuss in detail the Turing reaction-diffusion system, I will study a conceptually related and equally important system, which also couples local nonlinear oscillators via diffusive spatial motion: the "excitable medium" equations that model, among other systems, the propagation of nerve impulses and of electrical waves on the surface of the heart. Let me start with some background on the modeling of the heart as dynamical system.

Describing the human heart as a nonlinear oscillator (in the dynamical systems sense) has a surprisingly long history, dating back to pioneering work of van der Pol and van der Mark in the 1920s(25). In their original model, these authors ignored the spatial distribution of electrical excitation, and further simplified the problem so that only a single nonlinear oscillator was involved. The result was the (now) celebrated van der Pol equation,

$$\frac{d^2v}{dt^2} - \alpha(1-v^2)\frac{dv}{dt} + \omega^2 v = 0, \tag{10}$$

which we recognize as a simple harmonic oscillator – just set $\alpha = 0$ – modified by a nonlinear term proportional to $\frac{dv}{dt}$. Although this (over-) simplified model can describe only periodic oscillations of the heart ("normal" heartbeats) [1], extensions of the model to coupled ODEs and related systems can describe both the heart's normal beating and a large number of interesting arrhythmias, some perhaps reflecting chaotic behavior.(26) As discussed in these proceedings, exciting recent developments in chaotic control may provide novel therapeutic means of controlling cardiac arrhythmias.(27)

To describe the spatial propagation of electrical pulses through heart tissue, mathematical biologists often use a specific set of "excitable medium"

---

[1] I can not resist pointing out that "death," in which there is no beating at all, is (sadly) also a solution to this equation, although (in contrast to reality) it is *unstable*, in that small perturbations about "death" actually grow in time!

equations, the coupled partial differential equations known as the "FitzHugh-Nagumo" equations.(28) The two dependent variables in these equations are u, which represents the "propagating" variable (the electrical potential or chemical concentration), and v, which represents the "recovery" or "controller" variable (the opening of an ion channel or the concentration of a slowly diffusing chemical) and which exhibits a refractory period. In one spatial dimension, the equations read

$$\frac{\partial u}{\partial t} = \left(u - \frac{1}{3}u^3 - v\right)/\epsilon + D\frac{\partial^2 u}{\partial x^2}$$
$$\frac{\partial v}{\partial t} = \epsilon\left(u + \beta - \frac{1}{2}v\right) + \delta D\frac{\partial^2 u}{\partial x^2}$$
(11)

where typically $\delta \ll 1$ to model the very slow diffusion of v. The remaining parameters are ϵ, which measures the ratio of the excitability time to the refractory time, and β, which determines the threshold of excitation. The forms of the equations in two and three spatial dimensions involve the appropriate generalizations of the Laplacian operator, $\nabla \cdot \nabla$.

In intuitive terms, the behavior of these excitable medium equations is as follows. By "excitable" medium one means that while small local perturbations of the stationary state quickly decay, excitations larger than a certain threshold (related to the β parameter discussed above) produce a large change in the concentrations, and the concentrations then relax slowly to their values in the initial stationary state. Because of diffusion, these localized changes in concentration propagate outward from their points of creation as coherent structures.

Only a very brief qualitative description of the coherent structures contained in this class of excitable medium equations is possible here, for a proper description requires an entire volume.(29) First, starting in one spatial dimension, one finds (for certain parameter ranges) solitary wave pulses analogous to those shown in Figure 1 for the single variable Fisher's equation. These represent propagating excitation fronts.

In two spatial dimensions, one finds a richer variety of coherent structures. There are propagating plane wave fronts, generalizations of the fronts in one-dimensional case. But in addition there are "target" patterns, which as the name suggests, are made up of concentric rings emanating from a single point. Target patterns are initiated from local "pacemaker" regions in which above-threshold excitations are generated in a time-periodic manner and are strikingly visible, for instance, as alternating red and blue rings in the ferroin-catalyzed BZ reaction when it occurs in a wide, shallow dish, so that the system is effectively two dimensional.

One also finds rotating spiral patterns, as sketched in Figure 2. Importantly, the characteristics of these patterns – for instance, the wavelength of the coherent structure (the distance between successive loops in the spiral) and the motion of the tip of the spiral (the tip can be fixed or periodic in time, or undergo very irregular meandering(28)) reveal important details of the underlying chemical or biological mechanisms,(30) as a recent experimental study of spiral patterns in the transport of Ca^{2+} near the plasma membrane in oocytes strikingly demonstrates.(31),(32)

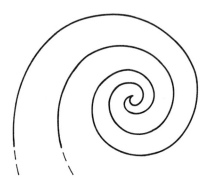

Figure 2 Sketch of a portion of a spiral pattern in the two-dimensional FitzHugh-Nagumo equation.

As a graphic illustration of the cross-disciplinary impact of the paradigm of coherent structures, Figure 3 compares spiral wave patterns observed in the BZ reaction with those observed in a biological system (aggregation of slime mold). The similarity is apparent.

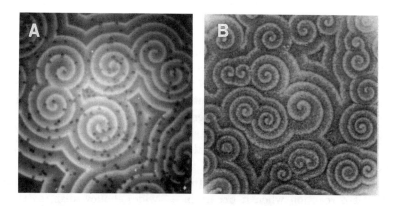

Figure 3 Spiral wave patterns in (A) the Belusov-Zhabotinsky reaction and (B) the aggregation of the slime mold *Dictyostelium discoideum*. (After

Epstein(31). Figure A is courtesy of A. Winfree (see (33)); Figure B is courtesy of P. C. Newell and F. Ross (see (34)).

Finally, in three dimensions, one finds yet another class of coherent structures – the "scroll waves" – which have the form sketched qualitatively in Figure 4. Serious readers are referred to Winfree's classic text(29) for a proper discussion of the implications of all these solutions for the behavior of normal and pathological cardiac tissue.

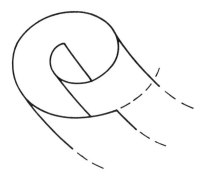

Figure 4 Sketch of a portion of a scroll wave in the three-dimensional FitzHugh-Nagumo equation.

APOLOGIA

I am afraid that, in attempting to introduce the concept of coherent structures in a very broad context in a very short article, I may have presented too many examples and not enough insight. In particular, I may have sought too much refuge in the mathematics, which while both concise and precise can be daunting for the non-specialist. But I hope that even a casual reader comes away convinced that coherent structures – solitons, fronts, and vortices – are an essential aspect of inherently nonlinear systems and that they have important implications for basic science and for technology. And I know that the dedicated reader will benefit from pursuing the ideas hinted at here in the many important and intriguing articles listed in the references.

REFERENCES

1. C. H. Tze, "Among the first texts to explain the 'soliton revolution'," *Physics Today* 35, 55-56, (June, 1982).
2. For a more thorough discussion (and illustrations!) of coherent structures, as well as other aspects of nonlinear science, presented in the spirit of the present article, see D. K. Campbell, "Nonlinear Science: From Paradigms to Practicalities," pp. 218-262 in *From Cardinals to Chaos* (Cambridge University Press, Cambridge, 1989).
3. A. R. Osborne and T. L. Burch, "Internal solitons in the Andaman Sea," *Science* 208, 451-460 (1980).
4. J. Wu, R. Keolian, and I. Rudnick, *Phys. Rev. Lett.* 52 1421 (1984); B. Denardo, W. Wright, S. Putterman, and A. Larraza, "Observation of a Kink Soliton on the Surface of a Liquid," *Phys. Rev. Lett.* 64, 1518-1521 (1990).
5. L. F. Mollenauer, J.P. Gordon, and M.N. Islam, "Soliton Propagation in long fibers with periodically compensated loss," *IEEE J. Quant. Elect.*, **QE-22**, 157-173 (1986); A. Hasegawa, "Numerical study of optical soliton transmission amplified periodically by stimulated Raman emission," *App. Opt.* 23, 3302-3309 (1984).
6. R. V. Coleman, B. Drake, P.K. Hansma, and G. Slough, "Charge-Density Waves Observed with a Tunneling Microscope," *Phys. Rev. Lett.* 55, 394-397 (1985).
7. N. H. Zabusky and M. D. Kruskal, "Interaction of "Solitons" in a Collisionless Plasma and Recurrence of Initial States," *Phys. Rev. Lett.* 15, 240-243 (1965).
8. D.J. Korteweg and J. DeVries, "On the change of form of long waves advancing in a rectangular canal and a new type of long stationary waves," *Philos. Mag.* 39, 422-443 (1895).
9. M. J. Ablowitz and H. A. Segur *Solitons and the Inverse Scattering Transform*, (SIAM, Philadelphia, 1981)
10. P. G. Drazin and R. S. John, *Solitons: An Introduction* (Cambridge Univ. Press, Cambridge, 1989).
11. A. C. Newell *Solitons in Mathematics and Physics*, (SIAM, Philadelphia, 1985).
12. M. Nakazawa, K. Suzuki, and E. Yamada, "20 Gbit/s, 1020Km, penalty-free soliton data transmission using erbium-doped fibre amplifiers," *Elect. Lett.* 28, 1046-1047 (1992)
13. A. J. Lotka, "Contribution to the theory of periodic reactions," *J. Phys. Chem.* 14, 271-279 (1910).
14. W.C. Bray, "A periodic reaction in homogeneous solution and its relation to catalysis," *J. Am. Chem. Soc.* 43, 1262-1267 (1921).
15. D. H. Shaw and H.C. Pritchard, "Homogeneous Periodic Reactions," *J. Am. Chem. Soc.* 72, 1403-1404 (1968).
16. R. J. Field and F. W. Schneider, "Oscillating Chemical Reactions and Nonlinear Dynamics," *J. Chem. Ed.* 66, 195-204 (1989).
17. I. R. Epstein, "The Role of Flow Systems in Far-from-Equilibrium Dynamics," *J. Chem. Ed.* 66, 191-195 (1989).
18. R. A. Fisher, "The Wave of Advance of an Advantageous Gene," *Ann. Eugenics* 7, 355-369 (1936).
19. M. J. Ablowitz and A. Zeppetella, "Explicit Solutions of Fisher's Equation for a Special Wave Speed," *Bull. Math. Biol.* 41, 835-840 (1979).
20. X. Y. Wang, "Exact and Explicit Solitary Wave Solutions for the Generalized Fisher Equation," *Phys. Lett A* 131, 277-279 (1988).
21. A. M. Turing, "The Chemical Basis of Morphogenesis," *Phil. Trans. R. Soc.* B 237, 37-72 (1952).

22. Q. Ouyang and H. L. Swinney, "Transition from a Uniform State to Hexagonal and Striped Turing Patterns," *Nature* **352**, 610-612 (1991).
23. V. Castets, E. Dulos, J. Boissonade, and P. De Kepper, "Experimental Evidence of a Sustained Standing Turing-Type Nonequilibrium Chemical Pattern," *Phys. Rev. Lett.* **64**, 2953-2956 (1990).
24. S. Kondo and R. Asai, "A Reaction-Diffusion wave on the skin of the marine angelfish *Pomacanthus*," *Nature* **376**, 765-768 (1995).
25. B. Van der Pol and J. Van der Mark, "The heartbeat considered as a relaxation oscillation and an electrical model of the heart," *Phil. Mag.* **6**, 763-775 (1928).
26. L. Glass, "Cardiac Arrhythmias and Circle Maps," *Chaos* **1**, 13-19 (1991).
27. A. Garfinkel, M. L. Spano, W. L. Ditto, and J. W. Weiss, "Controlling Cardiac Chaos," *Science* **257**, 1230-1235 (1992); F. X. Witkowski *et. al*, "Evidence for Determinism in Ventricular Fibrillation," *Phys. Rev. Lett.* **75**, 1230-1233 (1995).
28. For an excellent survey of these equations and additional references, see A. T. Winfree, "Varieties of spiral wave behavior: an experimentalist's approach to the theory of excitable media," *Chaos* **1**, 303-334 (1991).
29. A. T. Winfree, *When Time Breaks Down* (Princeton, 1987).
30. H. L. Swinney and V. I. Krinsky, *Waves and Patterns in Chemical and Biological Media* (North Holland, Amsterdam, 1991).
31. I. R. Epstein, "Spiral Waves in Chemistry and Biology," *Science* **252**, 67 (1991).
32. J. Lechleiter, S. Girard, E. Peralta, and D. Clapham, "Spiral Calcium Wave Propagation and Annihilation in *Xenopus laevis* Oocytes," *Science* **252**, 123-126 (1991).
33. A. T. Winfree and S. H. Strogatz, "Simple Chemical Waves in Three Dimensions. I: Geometrically simple waves," *Physica D* **8**, 35-49 (1983).
34. P. C. Newell and F. M. Ross, "Inhibition by Adenosine of the Aggregation Centre Initiation and Cyclic AMP Binding is *Dictyostelium*," *J. Gen. Microbio.* **128**, 2715-2724 (1982).

David K. Campbell received his B.A. degree in Chemistry and Physics summa cum laude from Harvard University in 1966 and was a Marshall Scholar and National Science Foundation Pre-Doctoral Fellow at Cambridge University, where he obtained his Ph.D. in 1970 in Theoretical Physics and Applied Mathematics. Following two post-doctoral fellowships at the University of Illinois and the Institute for Advanced Study in Princeton, he joined Los Alamos National laboratory as the first J. Robert Oppenheimer Fellow in 1974. During his nearly twenty-year tenure at Los Alamos, Dr. Campbell held a number of positions, including the Directorship of the Center of Nonlinear Studies (CNLS), a post he held from 1985-1992. In March of 1992 he assumed his present position as Professor and Head of the Department of Physics at the University of Illinois, Urbana-Champaign. His professional scientific interests are presently focused on several aspects of nonlinear phenomena, including general questions concerning the interaction of nonlinear waves and chaotic dynamics, applications of nonlinear models to real physical systems, particularly in the area of solid state physics, and development of pedagogical material concerning nonlinear processes. He is the founding and present Editor-in-Chief of the journal *CHAOS* and has served as a member of several national and international committees fostering nonlinear science. He is a Fellow of the American Physical Society and the American Association for the Advancement of Science, and has been honored with visiting research and distinguished lecturer positions in the U.S., as well as France, Germany, Russia, China, and Japan.

From Self-Organization to Emergence: Aesthetic Implications of Shifting Ideas of Organization

N. Katherine Hayles

English Department, University of California-Los Angeles
Los Angeles CA 90095-1530

Abstract. From 1945-95, a shift took place within cybernetics from a paradigm emphasizing self-organization to one emphasizing emergence. Central in bringing about this shift was the spread of the microcomputer. With its greatly enhanced processing speed and memory capabilities, the microcomputer made simulations possible that could not have been done before. The microcomputer has also been instrumental in effecting a similar change within literary texts. To exemplify the aesthetic implications of the shift from self- organization to emergence, the chapter discusses Vladmir Nabokov's *Pale Fire* and Milorad Pavić's *Dictionary of the Khazars: A Lexicon Novel in 100,000 Words*.

"To organize," according to the *American Heritage Dictionary*, means "to bring into existence formally," to "start, establish, create, constitute, set up, found, originate, institute." The definition highlights the foundational role that ideas about organization play in the analysis of complex systems. Because organization specifies how a system's components relate to each other, it can be considered a meta-concept, or a concept about concepts. For this reason, changes in ideas about organization are likely to be especially influential. A significant shift in what it means for a system to organize itself occurred from 1946 to the present. The shift can be conveniently traced through certain developments within cybernetics.[1] In brief, it consists of a movement from self-organization to emergence. In the following pages, I will detail some of the historical developments that went into creating this shift and outline consequences that followed from it. Important in delineating this shift are the key concepts of feedback loops, homeostasis, reflexivity, circular causality, and autopoiesis. These concepts did not, of course, exist in isolation. Also

---

[1] The shift is discussed in more detail in Hayles (13). Steve Heims has written a history of the Macy conferences (14), and P. R. Masani includes additional information in the context of Norbert Wiener's life (25). A collection of papers talking about self-organization as a paradigm shift appears in Wolfgang Krohn (17).

important are the material objects constructed to instantiate them and the historical and discursive contexts in which they were embedded. The concepts were intended to explain the behavior of complex systems. What complexity meant, however, was subject to changing interpretation. As we shall see, complexity in cybernetics was inextricably bound up with reflexivity. Within cybernetics, complex systems signified not only nonlinear dynamical behavior, but also systems that included the observer as part of their configuration.

In contrast to Michael Foucault (10), who has argued that a sharp epistemic break characterizes the movement from one period to another, my research into cybernetics suggests that change occurs through a process that I will call seriation (a term appropriated from a dating technique used in archeological anthropology). In seriation, change is incremental and variegated, overlapping old and new. Seriation can be seen in many historical processes. Consider the automobile: how does one think a car? By starting with an existing artifact, the horse carriage, and imagining that the carriage can move without the horses. Only gradually are the full implications of this shift realized through a progression of artifactual and conceptual changes. Greater speed necessitates windshields; still greater speed, aerodynamic styling. If one looks at two points far enough apart along a time line, the change may appear to be a sharp epistemic break of the kind Foucault described. But lying between those two points are a series of material and conceptual instantiations that constitute the microstructure of change. At the intermediate points, seriation unfolds through a progression of artifacts that partly innovate upon and partly replicate what came before.

The shift from self-organization to emergence can be understood in aesthetic as well as scientific terms. During the 1960s and 1970s, a group of literary texts appeared that were fascinated with the idea that texts, like complex physical systems, could engage in self-organization.[2] By the mid-1980s, the emphasis had shifted to emergence. Whereas texts evoking self-organization emphasized closure, stability, and the continual production and re-production of a system by itself, texts organized according to emergence engaged the reader in an upward spiral of increasing complexity. To illustrate self-organization, I will discuss the self-organizing tactics of the 1962 novel *Pale Fire* by Vladimir Nabokov, a Russian-American writer (31); to illustrate emergence, I will turn to the 1988 *Dictionary of the Khazars: A Lexicon Novel in 100,000 Words* by Milorad Pavić, a Serb-Croatian writer (34). A caveat is in order here. Whereas physical systems really can undergo self-organization, textual systems achieve organization (whether

---

[2]David Porush has written about some of these texts in the context of cybernetic fiction (35). William Paulson (33) talks more generally about how self-organization enters into literary reading and writing.

self-organization or emergence) only in conjunction with the mind of a reader. My conclusion will concern the relation between different ideas of organization and effects produced in and through the readers of the texts.

The Self-Organization of Self-Organization

A landmark in the history of self-organization is Heinz von Foerster's *Observing Systems*, a collection of essays written from 1960-1972 (40). Throughout these essays, von Foerster wrestles with how to integrate cybernetics with reflexivity. For our purposes, we can define reflexivity as broadening the scope of inquiry to include, in addition to cybernetic systems, the observers who constitute the systems as such. In cybernetics, the turn toward reflexivity has been called second-order cybernetics, or the cybernetics of cyberneticians. It was perhaps inevitable that reflexivity would become an important topic in cybernetics, for a central concept within the field is the feedback loop. Reflexivity enters the picture as soon as the loop is seen to run not only between components within the system, but also between the observer and the system. Foerster's punning title highlights the ambiguity. On the one hand, "observing systems" is what observers do; on the other, they are themselves systems that observe and that can therefore become objects of study to other observers. Von Foerster inherited the problem of reflexivity from the Macy conferences, a seminal series of annual meetings sponsored by the Josiah Macy Foundation and held from 1946-55. As co-editor of the transcripts from the Macy Conferences, von Foerster was very much aware of ideas about organization that were dominant during the Macy period (41). To understand how reflexivity entered into the mainstream of cybernetics, we can pick up the thread of our story with the discussions that went on at the Macy conferences.

During the Macy period, homeostasis was understood as the ability of a system to return to equilibrium after a disturbance. To demonstrate the concept, W. Ross Ashby designed the homeostat, a relatively simple electrical device constructed with variable resistors and transducers (2). The homeostat's goal was to adjust itself so that whenever a perturbation was introduced from the outside, it compensated to keep certain variables with pre-determined limits. It is clear from the Macy transcripts that the recent cataclysm of World War II was very much on the minds of the conferees. Ashby was not alone in equating homeostasis with the survival of individual organisms and, indeed, of entire societies. He argued that homeostasis had little value if it obtained only for systems in isolation; the more significant problem was how an organism could maintain equilibrium in the face of a chaotic environment. Accordingly, he designed his homeostat as four units which could be arranged in different configurations to represent an organism and its environment. For example, in one configuration unit 1 might be the organism, while units 2, 3, and 4 were the environment. Instantiated in the homeostat, then, is an idea of organization emphasizing equilibrium and the "circular causality" of feedback loops connecting system and environment.

Von Foerster realized circular causality's deeper implications; it could also connect the observer to the system he observes. In the early essays in *Observing Systems*, he remarks upon the importance of reflexivity, but he has no systematic way to conceptualize how the observer can be included so as to make a significant contribution to systems theory. In one essay, for example, he suggests that self-organization is a contradiction in terms, for it takes an observer to notice that a system is self-organized, in which case it is not self-organized but organized by an observer's perception (42). Elsewhere, he posits an infinite regress of an observer who is observed in turn by another, so on to infinity (42). Nevertheless, he also sees how powerful reflexivity can be in recontextualizing problems. Opposed to what he saw as the extreme reductionism of behaviorist experiments, he uses reflexivity to alter the experiment's meaning. From a behaviorist's point of view, animals and humans can be considered black boxes that give a predictable outcome for a given input; von Foerster's name for this kind of black box was a "trivial machine."[3] By contrast, a "non-trivial machine" produces output that cannot be predicted solely from its input. Rather than accept the behaviorist view, von Foerster suggests that the focus be shifted to the experimenter. The question then becomes how the experimenter, operating as a non-trivial machine, is able to alter his environment so as to make the objects of his experiments behave as if they were trivial machines. Broadening the scope of inquiry beyond the premises of behaviorist psychology, von Foerster uses reflexivity to insist that the behaviorist picture should not be understood as an objective reflection of the external world. Rather, it is constituted through a set of power relations that determines who gets to operate non-trivially, and who is forced into triviality.

Not until von Foerster encountered the theories of Humberto Maturana and Francisco Varela, however, did the idea of the observer enter in a systematically powerful way into his work.[4] One of the last essays in

---

[3]The argument appears in "Molecular Ethology: An Immodest Proposal for Semantic Clarification" (42, pp. 150-88).

[4]Maturana and Varela are the co-authors of *Autopoiesis* and *Cognition* (26) and *The Tree of Knowledge* (27). Varela was Maturana's student; Maturana began to develop the concept of autopoiesis before he teamed up with Varela, and parts of *Autopoiesis and Cognition* (26) are written under his name alone. *The Tree of Knowledge* (27) was the last work Varela and Maturana published together. After that, Varela began to take a different path in his work on autonomous systems and artificial life, specifically with regard to the role that the environment

Observing Systems, "Notes toward an Epistemology of Living Things," clearly shows Maturana's influence, in particular (42, pp. 257-272). The essay restates Maturana and Varela's ideas through a series of numbered statements that function as quasi-mathematical theorems, with the last coming back to the first to form a closed circle. The closed form is significant. Through his work on perceptual systems in frogs and primates, Maturana had become convinced that perception is not fundamentally representational (22, 28). He made the bold move of asserting that living systems are informationally closed to their environments. Starting from this premise, he developed a theory of living systems that he called autopoiesis, or self-making (26). The central insight of autopoiesis is to envision a reciprocal, tautological relation in which the components produce a system through their interactions with each other, and the system produces the components as components (rather than isolated parts) through its operation as a system. Thus the components produce the system, and the system produces the components. In envisioning this closed circle of interactions, Maturana shifted the earlier emphasis on feedback loops between the system and environment to feedback loops within the system.

Although Maturana and Varela modified homeostasis, they did not leave it altogether behind. The sense of dynamic stability that homeostasis conveys is reinscribed in the theory through the proposition that a living system, while capable of change through its structural coupling with the surrounding medium, nevertheless always maintains some quality unchanged. This quality they called organization, which they defined as the relationships instantiated in the interactions of a system's components with each other (26). An autopoietic system has as its one and only goal the continual production and re-production of its organization.[5] If its organization changes, it ceases to be that kind of system and becomes something else, for example dead rather than alive. All living systems, according to Maturana, are autopoietic. Indeed, he viewed autopoiesis as the necessary and sufficient condition for a system

---

plays in shaping an organism's response.

[5]*The Tree of Knowledge* (27), co-authored by Maturana and Varela, tries to integrate autopoiesis with evolution. They argue that the organism is always structurally coupled to its environment, so that if the nature of that structural coupling changes, the organism's structure must also change to maintain the coupling intact. Thus there is a "natural drift" in the organism's structure - yet its organization is alleged to be always the same. There are obviously some problems with this argument, since the organization of organisms does change over time as speciation occurs. The implicit contradictions point to the uneasy relation that autopoiesis, with its roots in homeostasis, has with a theory of change such as evolution. Paul Dell is more convincing in his argument that autopoiesis, unlike homeostasis, does not suppose a return to an equilibrium value and so is able to encompass change in a way that homeostasis is not (8).

to qualify as living.

The premise that living systems are informationally closed to their environment means, of course, that no information passes from the environment to the organism. Events in the surrounding medium trigger further events in the organism, but in Maturana and Varela's view, the interior events are not caused by the external incidents. Rather, they are determined by the organism's systemic organization and structure. Although this seems counter-intuitive, there are some ways in which it corresponds with everyday experience. A slap in the face may, for example, be associated with very different responses depending on whether it is done to a Prussian military officer or a yearning masochist. Why does the same event trigger different responses? Presumably because the two systems have different psychological (and hence physical) organizations. If one event does not cause another, it follows that causality is not an inherent attribute of the relation between system and environment but rather a "punctuation" added by an observer.

Although the idea that observers "punctuate" systems is consistent with, and indeed necessitated by, the premise of informational closure, it raises troubling questions about the role of the observer. On the one hand, Maturana and Varela posit a "domain of the observer" as a realm conceptually distinct from the systems it punctuates; on the other hand, the observer must herself be the product of autopoietic processes, since they are the basis for all life (26, 27). The observer thus occupies a dual role, positioned at once outside and inside the circular productions of an autopoietic system. This double positioning points toward a significant absence in the theory. Missing in Maturana and Varela's account is any detailed description of how cognitive processes can "bubble up" from autopoietic processes. Lacking such a description, the theory has no convincing way to connect the aconscious working of the autopoietic system with the cognition that, Maturana and Varela nevertheless insist, must somehow be one and the same as autopoietic processes.

The double positioning also indicates the seriated relationship between autopoiesis and its predecessor, homeostasis. In its attempt to posit a "domain of the observer" separate from the system, autopoiesis reinscribes the objectivist assumptions of the early Macy period, even as it calls them into question; in insisting that the observer is produced by autopoietic processes, the theory reinscribes circular causality, even as it calls that idea into question. These reinscriptions inevitably limit the extent to which the theory can effect far-reaching innovations. Just as it would be a mistake to underestimate the revolutionary implications of Maturana and Varela's theory, so it would be an error to see the theory as a complete break from what came before. The trajectory arcing from Ashby through von Foerster to Maturana shows substantial change, in that reflexivity becomes progressively more embedded in the conceptual structure of the theories. The trajectory also shows replication in the continuing importance of notions of closure and stability, which by the end of this period have been reconceptualized as a system's continuing self-production of itself.

Fast-forward to the 1980s, when the concept of emergence begins to take hold. Emergence differs from earlier ideas about self-organization because it focuses on a system's ability to evolve rather than to maintain itself in a steady state. The break can be seen in Luc Steels' distinction between first- and second-order emergence.[6] According to Steels, an artificial life researcher, first-order emergence occurs when a system exhibits properties not inherent in its components. Already we can see a shift between this idea and Maturana and Varela's concept of autopoiesis. Whereas autopoiesis envisions the system and its components as mutually producing each other in a dynamic that is symmetrical and tautological, first-order emergence implies that there exists an asymmetry between the local level of the components and the global level of the system. An even larger break occurs with second-order emergence, defined as an emergent property that bestows additional functionality on the system, especially a functionality that increases the system's ability to process information. Whereas first-order emergence is associated with a system's ability to evolve, second-order emergence is associated with its ability to evolve the capacity to evolve. In Richard Dawkins' apt phrase, second order emergence is concerned with the "evolution of evolvability."[7]

Significantly, much of the work on emergence has taken place in the context of computer simulation, either directly within the computer or in artificial life forms that rely on computerized components. Emergence is obviously much easier to conceptualize and model when one has a high-speed computer medium within which the processes can diversify, compete for resources, mutate and "breed."[8] It is thus no accident that the ability to process information is identified as a privileged form of emergence, for within

---

[6]Emergence has proved to be a difficult concept to define precisely; Steels develops an elaborate conceptual scheme to accomplish the task (40). Helpful comparisons of artificial life and artificial intelligence that speak to the shift toward emergence can be found in Dyer (9).

[7]Dawkins develops the concept through the "Blind Watchmaker" computer program, which he discusses in (6) and (7). Kauffman also talks about the evolution of the ability to evolve in (16).

[8]Thomas S. Ray's "Tierra" program illustrates how programs representing different species can "breed" in the computer (39). See also Christopher Langton's discussion of artificial life as a research program in (18).

the computer medium, superior information processing capabilities translate directly into higher reproductive fitness. It is also no accident that the shift from homeostasis to emergence took place from 1960 to 1990, the period when microcomputers were becoming widely accessible to researchers, computer memory and processing speed were increasing at extraordinary rates, and parallel processing architecture made possible a much more complex interplay between autonomous agents and interconnected programs than had been the case before.

Literature has also been affected by the computer revolution, particularly through the development of hypertext and hypermedia.[9] As many readers may know, hypertext is an electronic document that uses a link structure to connect blocks of text or lexias. In an electronic hypertext, certain words or icons that appear on the screen are interactive. When a reader clicks on them, the links are activated and new text appears on screen. Since it is the reader who decides when and where to activate the links, another way to think about hypertext is as a form of electronic writing that is reader-controlled. During the period from 1960 to 1995, hypertext developed from a twinkle in Ted Nelson's eye (32) to a full-fledged writing medium. The World Wide Web is a vast hypertext document, and most documents within it are also hypertexts. The explosion of hypertext is also apparent in literary productions. One form of literary hypertexts are interactive fictions in which the reader chooses the narrative path, more sophisticated versions of the "Choose Your Own Adventure" stories many of us read as children.[10] Others are scholarly editions, critical and interpretive works, and pedagogical texts. Literary theorists are scrambling to understand the full implications of the shift from a codex print book, whose pages are bound together into a fixed sequence, to an electronic hypertext, which exists as a large set of potential readings that readers can activate in different ways. Since the reader becomes a collaborator with the writer in bringing the narrative into existence, many critics are now using the term "wreader" to refer to this combined writer/reader function (15, 21). As the usage implies, hypertext forces a re-conceptualization of terms fundamental to literary criticism, including text, writer, and reader (15, 21).

Just as the computer was intimately involved in the transition from self-organization to emergence within cybernetics and related fields, so the

---

[9]The impact of the computer revolution on literature is discussed by J. D. Bolter (3), M. Joyce (15), R. Lanham (19), G. Landow (20) and a number of literary theorists (21).

[10]For a discussion of the different kinds of hypertexts now proliferating through the culture, from the aircraft industry to museum walk-throughs, see Landow (21).

computer also looms large in the transition from self-organizing strategies of *Pale Fire* to the complex emergences of *Dictionary of the Khazars*. Nabokov wrote *Pale Fire* standing at a podium and scribbling in a nearly illegible hand on a stack of small numbered note cards--much the same medium as he describes John Shade, the novel's protagonist, using to write his epic poem. Control was everything to Nabokov; it is unlikely that he would have been pleased by the prospect of a reader perusing his books in some other order than that which he, the author, had dictated.[11] *Dictionary of the Khazars*, by contrast, functions like a hypertext in emphasizing that there is no set order in which the novel must be read. The author suggests several reading protocols, all different, and then encourages the reader to experiment further on her own. It is no exaggeration to say that *Dictionary of the Khazars* is a print hypertext, an attempt to create within the print medium the complexities and indeterminacies characteristic of electronic hypertexts. At the same time, it is significant that Pavić glances back to print, and even further back to manuscript, to instantiate this vision.[12] On many levels, *Dictionary of the Khazars* insists that the past is not left behind but continues to echo through the future. Time, the narrator tells us in one of his characteristically enigmatic pronouncements, is the part of eternity that runs late. Seriation thus does more than help to form this text. It is also inscribed into the text's content, as the narrative emerges from the disjunctions and replications between conflicting voices speaking different stories at various times that are somehow all the same story.

In my description of these texts, I will show how differences in organization are instantiated in form, content, and reader response. These differences are consistent with the shift from self-organization to emergence traced earlier through cybernetics. At issue is not so much direct influence between cybernetics and literature, as a larger social change mediated through

---

[11] A discussion of Nabokov's writing habits can be found in Proffer (37). The issue of control comes up, among other places, in the context of films made from Nabokov's novel. Nabokov has one of his characters in the novel *Ada, or Ardor* say that he does not want any films made of his work, for he hates to feel that his artistry has been tainted by the smell of another hand. There is evidence that Nabokov shared his character's sentiments. See Appel (1) for a discussion of the film adaptations.

[12] The fact that Pavić chooses a print rather than electronic medium shows how the existence of a new technology can cause an older medium to be re-conceptualized, a point that Bolter (3) makes.

computer technologies and electronic textuality.[13]

The Self-Organizing Strategies of Pale Fire

Reflections are at the heart of *Pale Fire*'s self-organizing strategies. Reflections in a mirror; reflections in a window that seem to project a cozy, firelit room onto the dark snow where a spectator, unseen by those within, stands watching; reflections from a flashlight playing over a ghostly scene in an old barn or angling through an ancient passageway connecting a palace bedroom with a theater backstage; reflections between the lives of John Shade and Charles Kinbote, distinguished American poet and his (possibly deranged) commentator. Reflection is embodied in the novel's form. The first section (neglecting for a moment the Foreword) consists of a 999-line poem occupying 34 pages; the second section is a commentary on the poem that swells to a rambling 228 pages (31). Out of this commentary, which touches the poem only at odd moments and in odd ways, a plot emerges. Charles Kinbote has come to a picturesque New England college town to teach on a visiting appointment. His next-door neighbor, John Shade, is in the midst of composing a work written in heroic couplets, after the manner of Alexander Pope. To those around him, Kinbote is a social misfit, an awkward, strange man who is obviously star-struck with his famous neighbor. To himself, he is a man with a secret and a mission. The secret: he is the exiled King of Zembla, escaped from the palace where he was held prisoner by revolutionaries and then fled to America with the help of a secret network of Zemblan royalists. The mission: to persuade John Shade to immortalize Kinbote's country and his reign in a magnificent poem that will compensate for the slings and arrows with which outrageous fortune has seen fit to stick him. Whether Kinbote is in fact the exiled King he fancies himself to be, or a self-deluded but harmless oaf who has fashioned a glamorous fantasy to compensate for a dreary life, is a question the novel poses but does not answer.

Reflection emerges not only within the poem and commentary considered separately, but also in the relation between them. The commentary is broken into sections numbered according to the lines they putatively gloss, but these overt connections grow increasingly marginal as the commentary's own narrative line comes to the fore and gains momentum. At the same time, the poem and commentary come together in other ways through mirror

---

[13] The effects of the materiality of media of communication are discussed in the essays in Gumbrecht (12). Lanham (19), Bolter (3), and Joyce (15) emphasize the crucial role of the computer in forming new ideas about textuality.

symmetries between Shade and Kinbote. Shade is heterosexual, Kinbote homosexual; Shade is happily married, Kinbote remembers with shame the pain he caused his wife, appropriately titled the Duchess of Payn; Shade's name eerily reflects the name of the revolutionary group which ousts the king and vows to hunt him down, the Shadows; Shade wrestles in his poem with the suicide of his unlovely daughter, Hazel, a social outcast whose pain Kinbote understands in some ways better than he.[14]

There are also deeper connections. Haunted by his daughter's death, Shade wonders if there is an afterlife in which she can find the happiness that eluded her here. He thinks he might have glimpsed an answer when his heart stops suddenly during a lecture. While he lies unconscious, technically dead, he has a vision of a white fountain. Later he reads in a newspaper of a woman, also technically dead, who while unconscious had a vision of a white fountain. Seeking confirmation, he explores the correspondence, only to discover that the account contains a misprint: the woman saw a mountain, not a fountain. Although his initial reaction is ironic disappointment ("Life Everlasting!--based on a misprint" (31, p. 62, l. 803)), on reflection he decides that the mistake is itself a message: "not text, but texture; not the dream/but topsy-turvical coincidence,/Not flimsy nonsense, but a web of sense" (31, p. 63, ll. 808-810). As if in obscure recognition of the commentary that will eventually attach itself to his poem, he affirms, "Yes! It sufficed that I in life could find/Some kind of link-and-bobolink, some kind/Of correlated pattern in the game,/Plexed artistry" (31, p. 63, ll. 811-814).

Shade's life ends shortly thereafter, in an incident as ambiguously encoded as the rest of the work. He is killed either by Jack Grey, an escaped inmate who mistakes him for Judge Goldsmith, the man who sentenced Grey to prison; or by Gradus, an assassin sent by the Shadows to kill Kinbote but who misses and hits Shade instead. The poem, written on a stack of index cards and unfinished at line 999, is scooped up by Kinbote in the confusion, who finds to his horror that Shade has not been writing the glorious story of Zembla but a tale woven out of his own concerns. Like Shade, however, Kinbote decides to make the best of the pale fire that life has given him. He is determined to find in the texture, if not the text, of the poem the linked reflections of his own story, weaving his tale into its lines through a commentary that, together with the poem, comprises the "plexed artistry" of the novel as a whole. Then, in hiding as much from the executors of Shade's estate as from the presumed assassins, he prefaces the whole with a

---

[14]For a discussion of the extensive role that mirror opposites play in Nabokov's work, see J. W. Connolly (5) for the early novels and L. Maddox for the novels written in English (among them *Pale Fire*) (24). See G. Green (11) for an analysis of the role that mirror opposites played in Nabokov's own life, especially his relationship to his brother, who was a homosexual.

"Foreword" that provides the end to the tale, even as it also initiates its beginning.

Through these closures, the work becomes a self-organizing entity in which the parts produce the whole and the whole produces the parts. Kinbote's commentary obviously would not exist without the poem, but the poem (since it has fallen into Kinbote's hands) also can not be published without the commentary. Moreover, neither poem nor commentary is complete in itself. The novel is produced by their interactions, by the mirror reversals and uncanny reflections that connect them. Like Maturana and Varela's autopoietic system, *Pale Fire* continually produces and re-produces its own organization. The structure is hinted at in the "Word Golf" entry in the index Kinbote constructs. "Word Golf," we know from the poem, is one of Shade's favorite games. The entry cross-references Word Golf to lass; the entry on lass refers us to mass; mass to mars, mare, and male; male to Word Golf. In pursuing these entries, we learn the game (a move consists of changing one letter; the game is complete when one arrives at an antonym of the original word, the object being to reach this goal in the minimum number of plays). We also play the game through the entries that both define and constitute its circular reflective structure. On a larger scale, when the antonyms constituted by Shade's poem and Kinbote's commentary are juxtaposed, the self-organization of the novel is complete and the system achieves closure, becoming a stable autopoietic entity. Kinbote has achieved the immortality he sought in much the same sense that Shade found reassurance in his quest--by becoming part of a self-organizing whole.

The organization instantiated in *Pale Fire*, then, emphasizes reflection, closure, stability, and self-organization. Observers occupy a dual position, as they did in autopoiesis. Sometimes they are positioned outside, as when Kinbote spies on John and Sybil Shade through the couple's living room window; at other times the observer is positioned inside the system, as when Kinbote and Shade are produced as fully realized characters through their mirror relation to one another. As we shall see, a very different idea of organization is instantiated in *Dictionary of the Khazars*. Where *Pale Fire* emphasizes dyads, it creates triads; where *Pale Fire* produces closure, the *Dictionary* repeatedly rejects closure in favor of spirals of increasing complexity.

Emergent Textuality in Dictionary of the Khazars

At the center of *Dictionary of the Khazars* lies a mystery, or more accurately an erasure, as if a picture made of salt had been left out in the rain. The *Dictionary* purports to bring together under one cover everything that is known about the Khazars, specifically about the Khazar polemic (34).

The polemic came about as follows. The king of the Khazars, known as the Kaghan, had a dream and was not satisfied with his soothsayer's interpretation. So he decided to invite to his court representatives from the three major religions, Christian, Moslem, and Jewish. He, along with his people, would convert to the religion of the one who gave the best interpretation. The representatives came and argued before the Kaghan; one was rewarded with the promised conversions. Which representative triumphed is not clear, however, for within a few decades the Khazars were defeated and their culture annihilated. All that remains is commentary upon the event in the three religions, and each claims that its representative was the victor. The *Dictionary* purports to bring all of this commentary together in three color-coded books: red for Christian, green for Moslem, yellow for Jewish. Within each book, entries are arranged alphabetically. Thus the reader can conveniently peruse the material and decide for herself what happened.

But how is this perusal to take place? As the narrator points out in his introduction, reading preferences vary. Some may prefer to read across the grain of the text, following the first entry on "Ateh," for example, through all three texts. Others may wish to turn the pages sequentially, as in an ordinary book. Still others may jump around in a fashion typical of dictionary reading, following cross-references as their tastes dictate. In a dictionary such reading choices are usually not consequential, for each entry is presumed to be more or less independent of those around it. By contrast, in a narrative where events are connected with one another, reading order matters. "Pulp Fiction" would not be the same film if the events were narrated chronologically rather than out of sequence; different orders of narration amount to different stories. To complicate the situation further, the narrator reminds us that reading protocols within the three traditions vary. Whether one reads a page left to right or right to left, top to bottom or bottom to top, depends on the tradition one follows. The *Dictionary* is subtitled *A Lexicon Novel in 100,000 Words.* Syntactically, "lexicon" snuggles next to "novel," but semantically, a gap yawns between them vast enough to accommodate millions of different narratives, for those 100,000 words can be combined in a staggeringly large number of ways. As with electronic hypertexts, the *Dictionary* is not a single narrative but a practically infinite set of narratives. A given narrative comes into existence only in conjunction with a reader's choices.

Even these complexities do not exhaust the text's permutations, however. The narrator goes to considerable trouble to document the extraordinary course of events through which the *Dictionary* came into existence. Like everything concerning the Khazars, at the center of this explanation looms a large hole or erasure. If we piece together information taken from various places in the text and straighten out the chronology, this is the story that emerges. The three representatives, St. Cyril for the Christians, Ibn Kora for the Moslems, and Isaac Sangari for the Jews, contest at the Khazar polemic in 861 AD. Each representative has a chronicler who writes down his version of what happened: Methodius for the Christians, Al-Bakri for the Moslems, Halevi for the Jews. These records, which consist of manuscripts written in the three different languages and scripts, were collected in the late seventeenth century by various people, who were not necessarily operating together or collecting for the same reasons. In one way or another, all the manuscript collections are destroyed. However, Theoctist Nikolsky, a priest who has seen them and has a photographic memory, reconstructs them from memory and dictates them to the printer Joseph Daubmannus in 1691. In addition to 500 normal copies, the printer also composes one with a poison dye and encases it in a gold binding, making at the same time a companion silver copy. In the Inquisition, all copies are destroyed except these two. The silver copy is lost; the gold one becomes the property of Dorfmer family and is divided up in succeeding generations between different heirs. The last pages meet an ignominious end when an old man in the Dorfmer family uses them to scoop the fat from his daily soup. Bombarded by this welter of information, the reader may not notice that we are never told how the present edition comes into existence, since all copies of its supposed predecessor, the 1691 Daubmannus edition, are destroyed.

The technique is typical of Pavić's narrative strategy: the more we learn, the less we know. The text proceeds not as a single story, nor even as a sequence of stories, but as stories nested within stories within stories, the veracity of which are qualified in significant ways. For example, we are told a story about Princess Ateh's death but cautioned by the narrator, "It is known Princess Ateh never managed to die" (34, p. 23). Clearly, then, the story cannot be correct. Moreover, although it is told by the narrator, he does not claim it but attributes it instead to Daubmannus, the *Dictionary's* first printer, who related it not "as a story about how Princess Ateh actually died, but about how it could have happened had she been able to die at all" (34, p. 23). Having thus assured us that the story is untrue, the narrator asserts that it is nevertheless harmless, through a metaphor that makes a hair-raising turn: "Just as wine does not turn the hair gray, so this story cannot cause anyone harm" (34, p. 23). The story that follows is about how Princess Ateh died when she saw in two mirrors, one fast and one slow, the fatal letters from the proscribed Khazar alphabet painted on her eyelids to keep her safe while she slept. In this story that cannot harm us, letters kill. Such playful--and fatal--ambiguities are everywhere.

Consider as another example the title page that the narrator reproduces from the 1691 edition, which we come upon after we have finished the narrator's "Preliminary Notes to the Second, Reconstructed and Revised, Edition" and before we begin the three books that comprise the *Dictionary* proper. Below the facsimile page, with its Latin title and authentic-looking scrollwork, the reader finds these words added by the editor: "Title page from the original (destroyed) 1691 Daubmannus edition of *The Khazar Dictionary* (Reconstruction)" (34, p. 17). The original is destroyed but nevertheless reconstructed through a process about which we have no information, and with revisions whose nature and extent we can only guess. Clearly the page is meant to tweak the bibliographer's nose, as well as give him nightmares. As for literary critics, the narrator in the last line of his introduction assures us "they are like cuckolded husbands: always the last to find out . . ." (34, p.15).

Slowly, out of the multiple ambiguities and links deconstructing and connecting the different entries, something like a meta-narrative emerges. Before their conversion, the Khazars have their own myths about the relation between the natural and supernatural worlds. These myths are instantiated in the practice of the dream-hunters, who pursue through the dreams of others characters who remain the same from one person's dream to the next. These snatches of dreams are held to be moments of special illumination. Eventually, we learn that they are believed to be parts of the body of Adam Ruhani (or Adam Cadmon), the third mind in the world, the Adam-before-Adam, Adam's "angel ancestor" (34, p.165). Like Lucifer, Adam Ruhani forgot himself and rebelled. When he recovered his senses, he returned to heaven, but in the meantime he had slipped from the third to the tenth rung of heaven's ladder. Consequently he is doomed to drift back and forth between the third and tenth rungs, his huge body stretched to even more gargantuan proportions by the indeterminacy of his locale. The dream hunters strive to re-assemble that unimaginably vast body by compiling a book made up of the dream fragments they collect. We also learn that the Khazars envision time as space; the information is a clue that Adam Ruhani's body is time translated into spatial terms, or more accurately that portion of time with meaning for human beings, signified by its incarnation into a vast human-like body.

The antagonists to this centuries-long project are the demons, who are like God in that they are immortal, but like humans in that they cannot see all of time at once. They do not want the dream-hunters to assemble even a small part of Adam Ruhani's body, for the assemblage will bring humans closer to God and so diminish the edge that the demons have over them in the cosmic scheme of things. At crucial points, demons intervene to destroy the collected dream fragments, which are none other than the manuscripts that go into making the *Dictionary of the Khazars*. Thus the body of our book is also the book of Adam's body. Or rather, a small part of his body, or a revised reconstruction of a small part of it. The difficulties of making connections through these proliferating ambiguities point to the heroic nature of the project, which is nothing less than trying to make time make human sense. No wonder that the links are essentially infinite, that the origin of the text is so uncertain, that its status seems to change from moment to moment, that any conclusions we wrest from it are inextricably woven together with conflicting and contradictory cultural traditions. The text can have no set sequence because sequence, chronology, and time are precisely what is at issue in its construction.

The *Dictionary's* hypertextual form recalls Michael Joyce's observation that while we experience in time, we remember in space (15). Joyce uses the aphorism to argue that hypertext, by exploiting the two-and-a half dimensions created by imaging stacks of texts on a computer screen, is essentially a spatial rather than temporal writing medium. Because of hypertext's linking structure, it is also a medium particularly suited to the associational nature of human memory.[15] In hypertext, the space of memory is projected onto the space of writing to create a topographical surface of extraordinary complexity. By adopting a hypertext form, the *Dictionary* is able to exploit these complexities, even though it is written in a print medium. Taking the conversion of time into space as its central trope, it achieves incarnation in a hypertextual form that converts the time of reading into a spatial adventure proliferating endlessly across a variegated textual plane comprised of real and fake title pages, illuminated and unilluminated letters, and alphabetical entries that do not preserve the order of the original since they are written in a different language (not to mention being translated from Serb-Croatian into English) and dispersed into culturally- and color-encoded books that clash as noisily with one another as red does with yellow or green.

The striking differences in organization between *Pale Fire* and the *Dictionary* can be illustrated through the roles that mirror images and reversals play in the two texts. In *Pale Fire*, closure is achieved when an event (the poem) and the commentary upon it come together. However distorted Kinbote's narrative may be as an interpretation of the poem, and however it threatens to usurp its putative reason for being and take center stage itself, the reader nevertheless can access the poem directly and judge for herself what has happened. In the *Dictionary*, by contrast, the event (the Khazar polemic) is missing. The commentary is all that exists; the referent now comes into being through the words it supposedly spawns, rather than the other way around. We can envision the contrast by thinking of an object reflected in a mirror. Ordinarily, we are not confused about this situation; the object is real, the reflection is a play of light derived from this reality. In *Pale Fire*, the reflection becomes so vibrant and interesting that it begins to take on a reality of its own, eclipsing the object that it supposedly reflects. In *Dictionary*, the object has disappeared from view altogether. It must be reconstructed obliquely from its reflections in not one but hundreds of mirrors, all of which

---

[15]Most writers on hyptertext trace the origin of the idea to an article by Vannevar Bush (4). Bush argued that a textual system based on association rather than linear sequence would be much better suited to human memory, which, he argued, operates primarily through association.

have various kinds of distortions. Closure achieved by mirror opposites coming together to make a whole is no longer an option. Instead, the putative referent emerges from a proliferation of representations in exceedingly diverse and unreliable media, from the mirror of salt into which Ateh, the Khazar princess, peers, to the tattooed skin of a Khazar envoy on which is inscribed the history of his people. As he loses appendages to various accidents and greedy buyers, parts of the history are lost; eventually he is flayed and his skin is stretched and tanned, thus distorting the history in the same way that a Mercator projector distorts the globe when it is flattened into two dimensions. Even this artifact is lost with time, however, and now exists only in the words signifying it, which nevertheless aim, as their grand project, to recover the body of Adam Cadmon, which is also the body of time and history and hence includes as one of its parts the Khazar history, including the story of the envoy.

The loss of the referent in *Dictionary* and its problematic reconstruction through reflections is worked out in great detail. An important way in which this reconstruction takes place is through the *Dictionary's* distinctive use of mirror images. Whereas in *Pale Fire* opposites came together to form a closed, self-reflective autopoietic system, in *Dictionary* their joining is a traumatic event marked by both revelation and annihilation. Moreover, the moment of joining includes a third party who uses its energy to ricochet the story in a different direction, thereby initiating yet another spiral in the novel's increasing complexity. The technique is exemplified in the story of Avram Brankovich and Samuel Cohen. Brankovich begins having a series of coherent dreams in which the same person always appears. He believes it is his destiny to seek this person out and meet him. Meanwhile, the other person (who turns out to be Samuel Cohen) is having dreams about Brankovich. In fact, Cohen and Brankovich are dreaming each other. Brankovich sleeps only during the day, while Cohen sleeps at night; Brankovich mumbles in Spanish (Cohen's language) when he sleeps, while Cohen mutters in Polish (Brankovich's native tongue) during his nightly dreams. Brankovich has become interested in the Khazars because he is convinced that the sect of the dream hunters knows about the kind of dreams he is having. He therefore begins a collection of manuscripts and artifacts having to do with them, working through the Christian tradition he has inherited. Cohen, meanwhile, pursues his clues through Jewish tales and artifacts.

Into this dyad is interjected Yusef Masudi, who first appears on stage as Brankovich's valet and whose full story we do not learn until the final Green Book, the book of Moslem sources. Through other people's dreams, Masudi has caught glimpses of a man whom he eventually learns is Samuel Cohen. But he cannot see the man clearly; a third intervenes. He deduces that the third is also a man who dreams of Cohen, but directly rather than through the dreams of others. To locate this man, Masudi offers for sale material he has already collected about the Khazars, working through Moslem sources. Sure enough, the buyer is an agent for Brankovich, to whom Masudi then offers his services as a valet. Observing Brankovich dreaming, he becomes convinced that at the instant when Cohen and Brankovich meet, there will be an opportunity for him to seize the gleam of a different kind of truth from the edge of their two mutually reflecting surfaces.

The climactic moment arrives on a battlefield near the Danube, where Brankovich is speared through the chest by the Turkish pasha's soldiers while sleeping in front of his tent. The last thing he sees in life is Samuel Cohen charging toward him; the instant Brankovich dies, Cohen falls down into a coma, since there is now no one to dream his life (34, p. 57). The pasha orders his soldiers to kill Masudi as well, but Masudi is able to prolong his life by revealing that Cohen is only sleeping. He begs to live one day longer so that he can catch Cohen's dream of Brankovich's death. To understand the full significance of the request, the reader must wait until the story is re-told in the Yellow Book and then re-told again in the Green Book. From these multiple accounts, including another nest of stories about Masudi's quest and the three transformations it undergoes, there emerges a larger pattern of what it means to have a third term interposed between mirror opposites.

One of the ways that the novel signifies this emergence is by structuring its narratives in fractal-like patterns, so that self-similar complexities emerge at every scale level, from the smallest details on up to the vast cosmic struggle between the dream hunters and demons. In this fractal structure, links are made between similar forms across scales as well as within the same scale. The combination of vertical, horizontal, and diagonal links gives the emergent stories coherence, despite their extreme spatial dispersion. The meta-narrative about the demons and dream hunters, for example, is dispersed into dozens of lexias. So distributed is it, in fact, that it does not usually become apparent to readers, even professionally experienced ones, until a second, third, or fourth reading. To reference it, the critic would have to instance literally hundreds of bits of information across multiple scales and ask the reader to draw inferences from them--a work which itself would be likely to take a hypertext form.[16]

To illustrate how this fractally dispersed linking structure works, let us return to the insertion of Yusef Masudi into the dyad of Brankovich/Cohen. In the Green Book, we learn that Masudi's first occupation is a lute player. He leaves music behind, or thinks he does, when he begins catching dreams of Cohen and becomes a dream hunter, compiling accounts of his dreams to contribute to the *Khazar Dictionary*. One night while staying at an inn, he

---

[16]I am endebted to Joseph Tabbi in a private communication for illustrating how a critical interpretation of *Dictionary* could itself take hypertext form.

hears an especially difficult passage in a lute song being played with a fingering he does not recognize. When he tracks down the player and confronts him, the player admits he was able to devise the unusual fingering because he has been using eleven appendages--ten fingers plus his tail. The player thus acknowledges that he is a demon, allowing the reader to identify him with the lute player we have heard about in previous lexias and to connect Masudi with the demon-dream hunter struggle. The demon persuades Masudi that the worthwhile quest for him to pursue is not to contribute to the *Dictionary*, which can be taken as the book of life, but rather to learn more about death. This he can do by catching Cohen's dream at the moment of Brankovich's death (thus making horizontal links between the present Green Book, the Red Book where Brankovich's story is first told, and the Yellow Book where we learn the most about Cohen). Since Cohen must dream everything Brankovich experiences, he must also dream his death. If Masudi can catch that dream, he will gain information that no living person knows--what it feels like to die. Thus Masudi begins his third vocation, which is like his second vocation (the quest to collect luminous dreams) but with a different goal (now to learn about death rather than the Khazars). The insertion of this third vocation into the previous two means that Masudi will seek to insert himself between Cohen and Brankovich. It also explains why, when he dies, he goes to hell, for he has fallen victim to the demon's blandishments--information we are given at several other dispersed points. Through these links, Masudi's triadic quest itself becomes a form that is fractally similar to his insertion as a third term between Brankovich and Cohen.

This example could be multiplied by the hundreds or thousands, since such complex linking strategies and fractally similar forms appear at many points within each lexia. As with artificial life simulations, the emergence of larger global narratives from local lexias is made possible by the extremely high connectivity of this text. In a real sense, the text is inexhaustible. While the number of items to be permuted is finite (the 100,000 words), the combinations they spawn are, for human purposes, essentially infinite.[17] Clearly, the organization of this text points to a different goal than does the organization of *Pale Fire*. The difference is not a contrast between the simple and the complex--no one is likely to call *Pale Fire* simple--but between

---

[17]As Paul Harris pointed out to me, similar strategies can be found in the work of the Oulipo, a French group dedicated to the invention of "potential literature" by exploiting the combinatorics of language (see W. F. Motte for some examples, (29). One of the better known is Raymond Queneau's *One Hundred Million Million Poems* (38), a volume of "potential poems." Despite the fantastic numbers of poems alluded to in the title, the volume is only a few pages long. The million million poems come from randomly combining each of the lines in the volume with its companions to create "potential" sonnets.

different kinds of complexity. One belongs to the era when self-organization is at the cutting edge of ideas about organization, the other to a later period when self-organization has, in seriated fashion, shaded into emergence.

The Emergent Reader

Some fields that make extensive use of computer simulation are experiencing a trajectory similar to the one traced by the disappearing referent in *Pale Fire* and *Dictionary*. At first computer simulations were intended to model a referent which was clearly and obviously there, much as a mirror reflects a real-life object. As computer power increased and simulation techniques advanced, it became possible to use the space of simulation to perform experiments that could not be done, even in principle, on the real-life object. At this point the simulation, although it had not altogether ceased to refer to the real-life object, acquired an epistemological autonomy that made it less like a model and more like an independent arena for investigation.[18] The next phase of the trajectory can be seen in artificial life, where programs have been designed to evolve on their own in ways unanticipated by the programmer. This capability led some artificial life researchers (for example, Christopher Langton (18)) to assert that alife programs are not models of life but life itself. In this view, alife programs do not merely simulate carbon-based organisms but rather constitute a parallel track of evolutionary development in silicon. The reflection, in other words, steps out of the mirror and lives an independent life on its own. The next step is for the putative referent to disappear, available in reconstructed form only through its multiple and ambiguous reflections within the space of simulation – a possibility already envisioned in numerous cyberspace novels and films.

The cybernetic agenda to include the observer as part of the picture has thus gone through reflexivity into simulation. Not only does simulation create a reality into which the observer can imaginatively project herself; it also creates a reality that goes ahead on its own, independent of whether the observer is there or not. This trajectory has brought about a curious inversion in the positioning of the observer. For von Foerster and his contemporaries, it was a revolutionary move to insist that the observer is part of the picture, for it undermined the naive realism of traditional views. In the context of simulation technology, insisting that the observer is part of the picture is a

---

[18]Jacob Steen Muller, from the Danish Hydraulic Institute, made this point in discussing simulation techniques used to model turbulent flow around the stanchions for the new long-distance bridge now being built in Denmark, a crucial issue in determining how the bridge will affect water flow, and consequently salinity, in the North Sea (30).

conservative move, for it suggests that the emergent processes that make the simulation appear to be "alive" in fact depend on the observer's interpretation of them for efficacy. The deeper insight embedded in both these moves is the realization that the world, whether a real-life object or a simulation, exists for us only to the extent that we interact with it.

There remains the question of where to place in this picture the reader, as a special kind of observer. We might be tempted to say that in physical systems self-organization or emergence takes place *within* the system, whereas with textual systems they occur in the relationship *between* the system and reader. The situation cannot be adequately understood through such a simple dichotomy, however, for the observer is already implicated in the construction of the system as system. What does it mean for a physical system to self-organize if there is no intelligent observer to understand it as such, and to what extent does the observer constitute self-organization through her categories of analysis?[19] If interaction is crucial to bringing-the-world into-being-for-us, the observer must in some sense already be in the picture for the picture to exist in human terms.[20] Nevertheless, I think it is possible to make a meaningful distinction between physical systems that engage in spontaneously arising order, and textual systems that are humanly designed artifacts. The self-organization of a physical system is not designed specifically to interact with the observer. While the observer must interact with the system to register it as such, the self-organizing processes that order the system are

---

[19] Niklas Luhmann has pointed out the artificiality of including an observer in a separate "domain," as Maturana and Varela do. Instead of this ad hoc procedure, Luhmann makes the important innovation of having the system derive from the moment when an observer makes a distinction. Following Spencer-Brown, Luhmann argues that form, in its fundamental sense, always implies the existence of an inside and an outside, that is, one side of a distinction versus the other side. In the case of systems theory, the distinction is between system and environment. For a discussion of these ideas in the context of constructivism, see Luhmann (23).

[20] Obviously, where the boundaries of the "system" are seen to fall is part of the issue in determining whether the observer is part of the "system." This is why Luhmann's emphasis (23) on beginning with a distinction is so important. The distinction between system and environment is the foundational move that establishes the system as such. Since it is the observer who makes the distinction, the observer is always implicated in the construction of a system. If we draw the boundary lines in a different place so that this first observer is included in the system (which is the move made in the preceding sentence), this new boundary line itself is a distinction made by an observer. Hence it is always possible to re-draw boundaries to include the observer, but in order to do so, there must be a second observer who makes the distinction that will include the first observer. To my mind, this is a much more cogent formulation of the double positioning that Maturana and Varela (26) fall back on in imagining an observer who is at once inside and outside some "domain" within a pre-existing system.

not closely coordinated with those that occur with the observer. The Belousov-Zhabotinskii (B-Z) reaction, for example, proceeds in its characteristic patterns whether the observer has seen it once or a hundred times.[21] The organizing process within the B-Z reaction do not change in conjunction with the self-organizing psychological and physical processes that instantiate the observer's growing familiarity with the reaction. Textual systems like *Pale Fire* and *Dictionary*, by contrast, are linguistic technologies designed precisely to effect changes in those who read them. These changes, moreover, are progressive and cumulative. The more we read and understand them, the more our psychological processes are molded to the contours of their designs. The reader of *Pale Fire* participates in as well as observes the text's self-organizing dynamics, just as the reader of *Dictionary* participates in the dynamics of emergence.

This phenomenon helps to explain why readers continually go back to literary texts and read them in new ways. The psyches of a new generation of readers are formed by the texts and culture of a new era, so they bring to the interaction with older texts a different dynamical interface. At the same time, readers are also incrementally affected by the older texts, so they return to contemporary texts and read them differently, too. In this view, humans are permeable membranes through which texts circulate to affect each other as well as their host organisms. In the host organism writing this chapter, texts exemplifying self-organization and emergence have come together to suggest that the emergent lesson of the *Dictionary* may be correct. As long as there are people to read the bodies of texts and bodies to instantiate the texts of people, the past will always mingle with the future in the elusive, synthesizing, and fractally complex third term of the present.

REFERENCES

1. Appel, A. Jr., *Nabokov's Dark Cinema*, New York. Oxford University Press, 1974.
2. Ashby, W. R., "Homeostasis," *Cybernetics*, ed. H. von Foerster, Ninth Conference, 1953, pp. 73-108.
3. Bolter, J. D., *Writing Space: The Computer, Hypertext, and the History of Writing*, Hillsdale. Lawrence Erlbaum and Associates, 1991.
4. Bush, V., "As We May Think," *Atlantic Monthly* 176: 101-8 (July 1945).
5. Connolly, J. W., *Nabokov's Early Fiction: Patterns of Self and Other*, New York. Cambridge University Press, 1992.

---

[21]The B-Z reaction is a classic example of a self-organizing system. For a discussion of its dynamics, see Winfree and Strogatz (43). In *Order Out of Chaos* (36), Prigogine and Stengers discuss the philosophical implication of self-organization as a paradigm shift from Being to Becoming.

6. Dawkins, R., *The Blind Watchmaker: Why the Evidence of Evolution Reveals a Universe Without a Design*. New York. Norton, 1987.
7. _____, "The Evolution of Evolvability," C. Langton, ed., *Artificial Life*, pp. 201-220.
8. Dell, Paul F., "Beyond Homeostasis: Toward a Concept of Coherence," *Family Process* 21: 21-44 (March 1982).
9. Dyer, M. G., "Toward Synthesizing Artificial Neural Networks that Exhibit Cooperative Intelligent Behavior: Some Open Issues in Artificial Life," *Artificial Life* 1: 111-134 (Fall 1993/Winter 1994).
10. Foucault, Michel, *The Order of Things: An Archeology of the Human Sciences*. New York. Vintage Books, 1970.
11. Green, G., *Freud and Nabokov*. Lincoln. University of Nebraska Press, 1988.
12. Gumbrecht, H. L. and K. L. Pfeiffer, *Materialities of Communication*. Stanford. Stanford University Press, 1994.
13. Hayles, N. K., "Boundary Disputes: Homeostasis, Reflexivity, and the Foundations of Cybernetics," *Configurations*. 3: 441-467 (1994).
14. Heims, S. J., *The Cybernetics Group*. Cambridge. Massachusetts Institute of Technology Press, 1991.
15. Joyce, M., *Of Two Minds: Hypertext Pedagogy and Poetics*. Ann Arbor. University of Michigan Press, 1994.
16. Kauffman, S. A., *The Origins of Order: Self-Organization and Selection in Evolution*, New York. Oxford University Press, 1993.
17. Krohn, Wolfgang et al, *Self-Organization: Portrait of a Scientific Revolution*. Dordrecht. Kluwer Academic Publishers, 1990.
18. Langton, C., ed., *Artificial Life*, Vol. VI, Santa Fe Institute Studies in the Sciences of Complexity. Redwood City. Addison-Wesley, 1989.
19. Lanham, R., *The Electronic Word: Democracy, Technology, and the Arts*. Chicago. University of Chicago Press, 1993.
20. Landow, G. P., *Hypertext: The Convergence of Contemporary Critical Theory and Technology*. Baltimore. Johns Hopkins University Press, 1992.
21. Landow, G. P., *Hyper/Text/Theory*. Baltimore. Johns Hopkins University Press, 1994.
22. Lettvin, J. Y., H. R. Maturana, W. S. McCulloch, and W. H. Pitts, "What the Frog's Eye Tells the Frog's Brain," *Proceedings of the Institute of Radio Engineers*. 47: 1940-1957 (1959).
23. Luhmann, N., "The Cognitive Program of Constructivism and a Reality That Remains Unknown," *Self-Organization: Portrait of a Scientific Revolution*, ed. Wolfgang Krohn et al., Dordrecht. Kluwer Academic Publishers, 1990, pp. 64-85.
24. Maddox, Lucy, *Nabokov's Novels in English*. Athens. University of Georgia Press, 1983.
25. Masani, P. R., *Norbert Wiener*. 1884-1964. Basel and Boston. Birkhuaser, 1989.
26. Maturana, H. R. and F. J. Varela, Autopoiesis and Cognition: The Realization of the Living. *Boston Studies in the Philosophy of Science*. 42: Dordrecht. Reidel, 1980.
27. _____, *The Tree of Knowledge*. Boston. New Science Library, 1987.
28. Maturana, H. R., G. Uribe, and S. Frenk, "A Biological Theory of Relativistic Color Coding in the Primate Retina," *Archivos de biologia y medicina experimentales*, Suppl. 1. Santiago Chile. 1969.
29. Motte, W. F., ed., *Oulipo: A Primer of Potential Literature*. Lincoln. University of Nebraska Press, 1986.
30. Muller, J. S., "Virtual Water: Modeling Rivers and Seas," "Virtual Nature" Conference,

Odense, Denmark. June 22-23, 1995.
31. Nabokov, Vladimir, *Pale Fire: A Novel*. New York. Putnam, 1962.
32. Nelson, Ted, *Literary Machines*. Sausalito, Calif. Mindful Press, 1990.
33. Paulson, William, *The Noise of Culture: Literary Texts in an Age of Information*. Ithaca. Cornell University Press, 1988.
34. Pavić, Milorad, *Dictionary of the Khazars: A Lexicon Novel in 100,000 Words*, translated by Christina Pribicevic-Zoric'. New York. Vintage Books, 1989.
35. Porush, David, *The Soft Machine: Cybernetic Fiction*. New York. Methuen, 1985.
36. Prigogine, I., and I. Stengers, *Order Out of Chaos: Man's New Dialogue with Nature*. New York. Bantam Books, 1984.
37. Proffer, C. R., *A Book of Things About Vladimir Nabokov*. Ann Arbor. Ardis, 1974.
38. Queneau, R., *One Hundred Million Million Poems*, English version by John Crombie, Paris. Kickshaws, 1983.
39. Ray, T. S., "An Approach to the Synthesis of Life," *Artificial Life II*, ed. C. G. Langton, C. Taylor, J. D. Farmer, and S. Rasmussen, Santa Fe Institute Studies in the Sciences of Complexity, vol. X, Redwood City. Addison-Wesley, 1991, pp. 371-407.
40. Steels, L., "The Artificial Life Roots of Artificial Intelligence," *Artificial Life* 1: 75-110 1994.
41. von Foerster, H., ed., *Cybernetics: Circular Causality and Feedback Mechanisms in Biological and Social Systems*, vols. 6-10, New York. Macy Foundation, 1949-55.
42. _____, *Observing Systems*, 2nd ed., Salinas: Intersystems Publications, 1984.
43. Winfree, A. T. and S. H. Strogatz, "Organizing Centers for Three-Dimensional Chemical Waves," *Nature* 311: 611-15 (1985).

N. Katherine Hayles is a Professor of English at UCLA since 1992 and holds advanced degrees in Chemistry, as well as English. She was a Professor of English at the University of Iowa from 1985-92. She received her B.S. degree in chemistry from Rochester Institute of Technology (with highest honors) in 1966, her M.S. degree in Chemistry from the California Institute of Technology in 1969, her M.A. degree in English Literature from Michigan State University in 1970, and her Ph.D. in English Literature from the University of Rochester in 1977. She has worked as a chemical consultant at Xerox and Beckman Instrument Company before receiving her Ph.D. She works in the area of literature and science. She is the author of several books, including *Chaos Bound: Orderly Disorder in Contemporary Literature and Science*, Cornell Univ. Press, 1990; *Chaos and Order: Complex Dynamics in Literature and Science*, University of Chicago Press, 1991; *The Cosmic Web: Scientific Field Models and Literary Strategies in the Twentieth Century,* Cornell Univ. Press, 1984. She is currently completing a book on the impact of information technologies on literature entitled *Virtual Bodies: Cybernetics, Literature, Information.*

Two Examples of Chaotic Dynamics in Fluids

Michael Gorman

Department of Physics
University of Houston
Houston, TX 77204-5506

Abstract. The dynamics of two experimental fluid systems will be discussed in terms of the ideas of chaos and nonlinear dynamics: the flow in a convection loop and the motion of a circular premixed flame. Each experimental system is described by a relatively simple mathematical model whose properties are well-known. These two examples demonstrate how descriptions of the dynamics in phase space manifest themselves as motions in physical space. They also illustrate the difficulties in implementing the mathematical techniques of analysis in the context of an experiment on a spatially extended system. These results are discussed in the context of applications to the study of similar phenomena in medicine.

Introduction

The popular book, Chaos[1], by James Gleick was a major event in the development of nonlinear science. First, it made the intellectual case that developments in a number of seemingly disparate areas in physical science, social science and mathematics were part of a large, coherent set of ideas which presented a new and different description of the physical world. Second, it presented these ideas at a level accessible to science graduates and medical researchers who did not have a sophisticated understanding of math or physics. Researchers throughout the sciences attempted to analyze their data in terms of these new ideas. However, the implementation of these new ideas in the context of physical systems proved considerably more difficult than Gleick's book suggested.

I have often wondered why the readers of Gleick's book do not go out and flap their arms at the approach of some undesirable weather pattern. After all, the "butterfly effect", as presented, says that a butterfly which flaps its wings in Beijing can affect the weather in the United States. You don't have to go to Beijing to flap your wings and try to affect the course of an approaching front. Another example concerns fractals. One of the most important characteristics of Mandlebrodt's description of fractals is that their structure is "the same" on different length scales.

He then goes on to show that many everyday physical phenomena have a fractal structure, including the coastline of the United States. Why then, one might ask, does the coastline of the United States not look the same on different length scales? One should be able to see "little" versions of the Eastern seaboard up and down the East Coast. These simple questions illustrate both the difficulty in applying the complex mathematical ideas underpinning chaos and fractals and the importance of having simple physical systems, as opposed to mathematical equations, as examples which demonstrate how to implement these techniques.

My research program was designed to bridge the gap between the theoretical ideas arising from the mathematics of chaotic systems and the analysis of laboratory experiments on dynamics in fluids (spatially extended systems). When I started my research group in 1981, I went in the opposite direction of most researchers in fluid dynamics. I decided to study simpler fluid systems than Taylor-Couette flow or Rayleigh-Benard convection, not more complicated ones. I stayed away from three-dimensional flows because it was my judgment that the dynamics got too complicated too fast. It defied my physical intuition that the analytical tools developed to analyze the chaotic dynamics arising from coupled sets of ODE's could describe the dynamics of fluid flows having a complex spatiotemporal structure described by PDE's.

This paper describes two experimental systems which exhibit chaotic dynamics throughout their parameter space. In section 1 the convection look is presented, an essentially one-dimensional fluid flow that is described by the Lorenz-like set of ordinary differential equations. In section 2 the dynamics of a premixed flame stabilized on a porous plug burner is presented. This is an essentially two-dimensional system which, over part of its parameter range, is described by the Kuramoto-Sivashinsky equation. The analysis of the dynamics observed in these two systems illustrates the differences between the numerical studies of the equations and the measured properties of the "corresponding" physical systems.

One of the liberties which a workshop such as this one provides is that one can give one's opinions on the important issues relevant to the topic. The purpose of the last two sections is to provide perspectives by someone who conducts experiments on idealized systems, which are not available in the literature. In section 3 I describe my assessment of the important factors in the application of the ideas of chaos and nonlinear dynamics to problems in medical science. And in section 4 I give my views of the historical context in which the research in chaotic dynamics is conducted.

Chaotic Dynamics of a Convection Loop

The flow in a convection loop is a particularly accessible example in which the chaotic dynamics associated with the physical motion of the fluid in the loop can be directly compared with the motion of the system point on attractors in phase space. A convection loop is a toroidal loop of fluid heated at constant flux over the bottom half and cooled at constant temperature over the top half. The industrial version of this device is called a thermosyphon. It is principally used in the emergency core cooling systems of nuclear power plants and in solar water heaters[2-4].

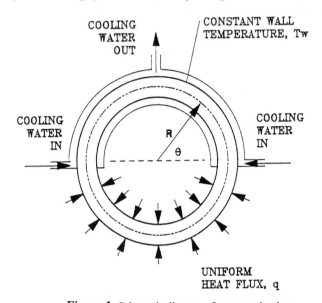

Figure 1. Schematic diagram of a convection loop

The three important dynamical variables are the sine and cosine Fourier coefficients of the temperature and the flow velocity. The dynamics can be studied by solving the equations which describe the flow or by examining the nature of the attractors in phase space. Each of the variables corresponds to an axis in this space. The dynamics is described by the motion of the system point in this space.

When the external heat flux is not present, there is no flow and no temperature differences in the fluid. The system point is a fixed point at the origin. At low values of the heat flux there is convection but no flow. At a critical value of the heat flux flow occurs in the clockwise or counterclockwise direction. There are now two stable fixed points corresponding to the two directions of flow and two different signs of the temperature difference as shown in Figure 1. Because these experiments are typically conducted by slow increases in the heat flux, the accessible initial conditions are very near the fixed points. As the heat flux is increased, the fixed points move further away from the origin.

At a second critical value of the heat flux, unbeknowst to the experimenter, the phase space has divided into two "basins" of attraction. The regions around the fixed points are encircled by cylindrical "tubes" which are unstable periodic orbits. Dynamics inside these tubes spirals into the fixed point. Dynamics outside these tubes is chaotic. As the external heat flux is slowly increased, it is difficult to perturb the experimental system enough to knock it out of these tubes and observe this chaotic behavior. Figure 2 shows a trajectory representative of this regime. The abrupt application of the heat flux moves the system point up the z axis. The flow begins in one direction but then reverses and oscillates with decreasing amplitude to a constant value in the opposite direction.

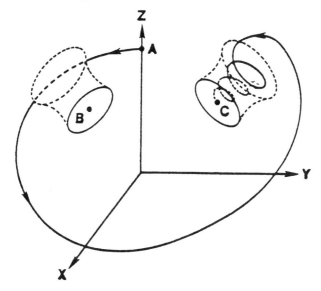

Figure 2. Diagram of trajectory in phase space.

At a third critical value of the heat flux, the fixed points become unstable and there is a single attractor, the chaotic attractor. All orbits are now chaotic. In physical terms an element of fluid now travels through the top half of the loop without being completely cooled when it reenters the heated section. The flow oscillates with increasing amplitude until it reverses direction. In phase space these oscillations correspond to circles of increasing size around the fixed points. The oscillations again build up, and similar reversals occur at irregular intervals. Figure 3 shows a time series of the temperature difference between two points in the loop and illustrates the irregular reversals.

The point in parameter space at which the fixed points become unstable is a subcritical Hopf bifurcation (SHB). In the experiment if the driving parameter (the applied heat flux) is now (slowly) lowered below this point, the flow will continue to be chaotic. In the phase space diagram tubes around the fixed points now reappear (because we are now below the SHB again), and the system remains outside those tubes. In this situation there are two separate basins of attraction--inside the barrier tubes around the fixed points and the rest of phase space which contains the chaotic attractor.

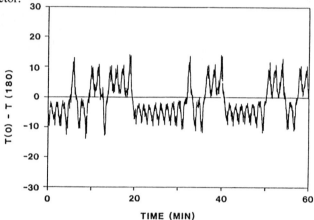

Figure 3. A representative time series of the measured values of temperature differences for flow in chaotic regime.

A somewhat different situation occurs if the heat flux is changed *abruptly* and reduced in magnitude so that the system is significantly below the Hopf bifurcation. The change in the driving parameter can be thought of as if the barrier tubes appeared suddenly. If the system point is (caught) inside the barrier tubes, it quickly spirals into the fixed point. In the experiment the oscillations of the chaotic flow regime, instead of increasing in amplitude, decrease in amplitude to steady flow. If the system point is (caught) outside the barrier tubes, it remains outside and the flow continues to be chaotic. In the experiment the oscillations continue to increase in amplitude and the flow remains chaotic.

This example illustrates how a simple fluid flow can have very complicated flow regimes which depend crucially on how the experiment is conducted. A knowledge of nonlinear dynamics is essential to understanding the flow regimes. The full range of possible observations is discussed in Reference 3.

Chaotic Dynamics of Premixed Flames

A premixed flame is a propagating front of chemical reaction. A flame front is stable when the burning velocity of the front equals the flow velocity of the gases. A burner causes heat loss and thereby reduces the velocity of the freely propagating front from the adiabatic flame speed to the velocity of the gases exiting the burner. In a Bunsen flame the inner cone is conical because the velocity profile of the premixed gases is approximately parabolic. A porous plug is used to create a uniform velocity profile which produces a flat flame front as shown in Figure 4. The burner is placed inside a combustion chamber made from process glass pipe. The flame motion is recorded using a Silicon Intensified Target Camera which views the flame through a mirror at the top of the chamber shown in Figure 5. Another arrangement uses a CCD camera mounted on top of the chamber and views the flame directly. A steady flame appears as a 3" luminous disk.

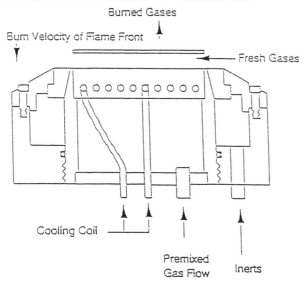

Figure 4. Schematic diagram of the porous plug burner.

The principal mechanism which makes the flame front unstable is the thermodiffusive instability. The Lewis number, the ratio of the thermal diffusivity to mass diffusivity, is the important parameter. Above a critical value, thermal diffusivity dominates and the flame front pulsates in periodic and chaotic motions. Below a (second) critical value, mass diffusivity dominates and the flame front physically curves away from the burner surface, forming brighter, hotter cells whose boundaries are demarked by darker, cooler cusps and folds[5]. Figure 6a shows an example of an ordered state of cellular flames[6]. The structure of the cusps and folds are not visible in this view because of the limited dynamic range of videotape.

Figure 6e shows a top and side view of the ratcheting mode in which a pattern of cells slowly drifts (~1 deg/sec) around a circular path[7]. The low angle side view gives a picture of the cusps and the folds. At most angles the bright troughs in the background shine through the dark cusps in the foreground. There is an optical illusion associated with this view. Most observers see waterdrops on the underside of a plate. The correct view is that the dark cusps and folds point away from the burner (at the bottom of the frame, not shown).

The experiments are typically conducted at a pressure of 1/2 atmosphere. As the flow rate and equivalence ratio are varied, the ordered states of concentric rings of cells bifurcate to other states. Figure 6b shows four sequential frames of videotape which depict a rotating state in which an outer ring of six cells rotates around an inner ring of two counterrotating cells[8]. Figure 6c shows a single central cell which assumes a sprial-like shape as it rotates[8]. Figure 6d shows a hopping motion in which the three cells sequentially change their angular position in the inner ring[9]. Figure 6e shows top and side views of ratcheting motion in which a pattern of (nineteen) 12/6/1 cells slowly drifts (~1 deg/sec) in a circular motion[7]. Figure 6f

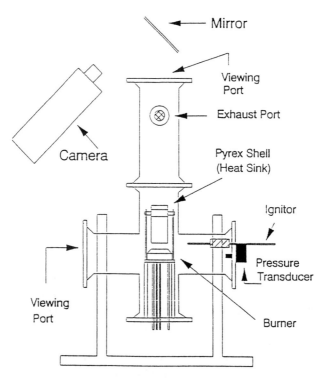

Figure 5. Schematic diagram of the combustion chamber.

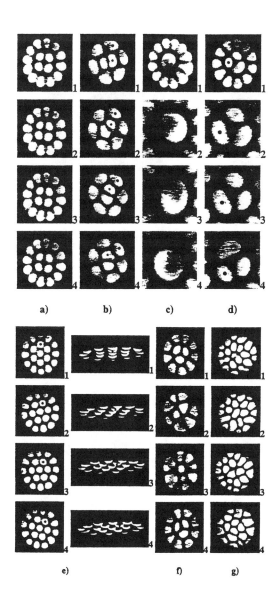

Figure 6. Four sequential frames of representative examples of dynamics of cellular flames: a) ordered states; b) rotating states; c) spiral state; d) hopping state; e) ratcheting state; f) intermittently ordered state and g) disordered state.

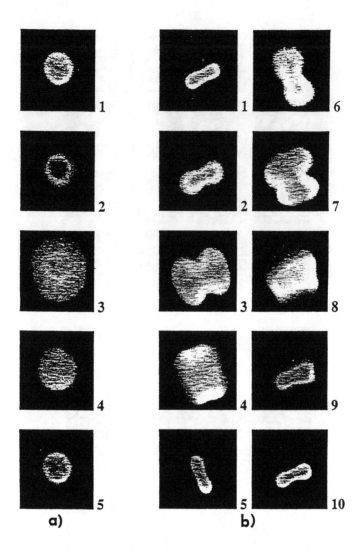

Figure 7. Sequential frames of videotape of a) the periodic radial mode and b) the periodic period-doubled radial mode.

shows an intermittently ordered state (frame 1) in which an ordered state of three inner cells and nine outer cells evolve into a highly irregular structure (frames 2 and 3) from which another ordered state abruptly appears at irregular intervals[10]. Figure 6g shows a disordered state in which the cells have irregular shapes and move around in a liquid-like manner[10]. These states are representative examples of the kinds of states observed in isobutane-air and propane-air cellular flames. References 6, 8-10 provide a more complete discussion of the characteristics of these states.

Figure 7a shows a representative example of a pulsating flame. In this radial state the flame front periodically changes its radial extent on the burner. As the flow parameters are adjusted near the extinction limit, the flame front loses its circular symmetry and assumes a dogbone-like shape as shown in Figure 7b. On successive minima this shape assumes orientations at (approximately) right angles to each other. The motion repeats every other cycle; the system has period-doubled. As the driving parameter is moved closer to the extinction limit, the motion becomes chaotic as sequential minima move about the burner surface[11].

Mpeg movies of these modes are available at the WWW site: http://vip.cs.utsa.edu/flames/overview.html.

Figure 7b and 6f capture the essential features of the chaotic dynamics of pulsating and cellular flames. In pulsating flames the correlation length of the dynamics extends over the entire burner, the power spectrum falls off exponentially, and the techniques of low-dimensional chaos can be used to analyze the dynamics. In cellular flames the correlation length is on the order of the cell size, the power spectrum falls off in a complicated manner, and the techniques of low-dimensional chaos cannot be used to analyze the dynamics. We think these results, which have been outlined in a series of papers in the combustion literature[6-13], are more general than these particular experiments. More extensive explanations and justifications will be presented in a series of papers in the physics literature.

Recommendations for Applications of Chaos in Medical Research

The two areas which have been thought to be excellent prospects for important (and lucrative) applications of chaos research are financial forecasting and irregular medical phenomena. Both have attracted lots of press. Both have failed to produce convincing results. Both rely on the "predictive" powers of chaotic dynamics.

I recently attended a workshop entitled, "A Critical Eye Towards Chaos", which tried to examine the prospects for potential applications of chaos research. A researcher who had done extensive work on chaotic phenomena in medical applications gave a presentation in which he said that 95% of the dimension calculations for chaos in medical systems were wrong. In a recent paper by Jim

Collins[14] there is a statement that there have been NO convincing identifications of chaos in medical systems.

The results of our research would suggest the following interpretation of these two assertions. It is my judgment that almost all of the nonperiodic medical phenomena are chaotic. Chaos is nonperiodic dynamics intrinsic to the system. Noise is nonperiodic dynamics extrinsic to the system.

However, the techniques of low-dimensional chaotic dynamics cannot be used for analysis. These systems are not low-dimensional; there is significant spatial variation in the dynamics. In that regard our view is that 100% of the dimension calculations in medical systems are wrong, just as Jim Collins' statement implies. But I would argue that the systems are chaotic; they just need more sophisticated analytical techniques.

The State of Texas supplies $1 billion to fund a laboratory for nonperiodic medical phenomena; it is called the Texas Medical Center (TMC). In my fifteen years at the University of Houston I have seriously engaged five or six medical researchers from the TMC about projects to investigate the chaotic aspects of a medical phenomena--from opthamologists (eyes) to cardiologists (heart) to orthorhynolarynchologists (posture), even to a study of dying from a systems perspective. For instance, 3500 Holter tapes (24 hr EKG's) of babies (who are ideal candidates for study because of the lack of complications from other factors) with arrhythmias go unanalyzed each year. I have assembled a team of physicians, biomedical engineers, theoretical and experimental physicists and mathematicians, but we have yet to secure any support.

Here are my suggestions for any project investigating nonperiodic medical phenomena from the perspective of chaos:

First, the most important tactic is finding the right problem, one that exhibits particularly simple characteristics which can be analyzed using the current techniques. Chaos is NOT a hammer to be used to pound on every nonperiodic medical phenomenon which comes along. The most important aspect of this paper is the opportunity to encourage physicists in medicine to be on the lookout for good problems. Physicists are trained to identify problems with simple, general characteristics. It usually takes a considerable amount of time to become acquainted with the particular aspects of the problem. One cannot underestimate the importance of day-to-day contact of the medical environment in developing a problem.

Second, it is absolutely crucial that researchers (most probably physicists) grounded in the application of techniques of chaotic dynamics to real physical systems be included as part of any research team. It is extraordinarily difficult to go from mathematical algorithms developed for the analysis of ODE's to nonideal experimental systems such as a human body which is subject to host of outside factors. The literature on chaos suggests exactly the reverse.

My suggestion is a team of mathematicians, physicists and medical researchers. Such collaborations are much easier today given the ease with which data can be transferred. The model for such a group is the group in which Bill Ditto works. It may seem implausible, but it is, in fact, true that the control of the dynamics of a magnetoelastic ribbon is more closely related to the control of arrythmias than many of the previous attempts by medical researchers.

A Version of the Past-- a View of the Present--a Vision of the Future

A few years ago I was at the APS sections and topical groups meeting because I was chairman of the Texas section. At lunch, another physicist sat down next to me and asked what kind of research I did. I said I studied the chaotic dynamics of fluids. He replied forcefully, "When are you guys going to do something useful?" In reply I mumbled something about medical applications looking promising and tried to emphasize the embryonic stage of the entire chaos research effort. Trying to change the subject, I asked him what his research effort was. He replied, "Plasma physics". Apoplectic barely does justice to a description of my response. Some people have no shame.

Where do nonlinear dynamics and chaos fit into the historical framework of physics? If they are part of a revolution,"like" quantum mechanics, as Mr. Gleick has argued, does that designation not imply a similar round of great technological progress which accompanied other revolutions in physics? If so, where are the applications? After all, it has been twenty years since the paper by Gollub and Swinney[15], which can reasonably be said to have begun the modern era of chaos research.

At the end of the last century there was a widely held opinion that physics was done, finished. The triumph of Maxwell in explaining electricity, magnetism and light in a unified framework complemented the accomplishments of a century before by Newton in the description of mechanical motion. The classical world was unified and complete.

Only a few subtleties remained to be resolved: some minor discrepancies between theory and experiment on black body radiation, some emissions of electric charge when a metal surface was exposed to ultraviolet light, an apparent inconsistency in calculating the force on a moving charge in magnetic field when the same question was considered from a reference frame moving with the charge. Nothing very important.

For the last fifty years there has been a sense that physics would last forever in its mid-century form, that there were an unlimited number of phenomena to investigate, explain or apply using quantum mechanics. Now the opposite seems true. The demise of the SSC was only the most obvious symbol of the recent stagnation.

There is tangible evidence that the technological influence of physics, 20th century physics, has peaked. The four great institutions which employed physicists in substantial numbers and which relied on innovations in physics for their commercial success are in decline, or, at least, in major retrenchment. IBM, AT&T, Xerox and the national laboratories have been destabilized, not by new companies run by physicists, but rather by nonphysicists--by a Harvard dropout (Microsoft), by a businessman (MCI), by two guys in a garage (Apple which took the concept of the Xerox Star and turned it into the Macintosh) and by a political leader (Gorbachev--who ended the cold war), respectively. Universities, which provide the physicists trained with technological skills, now find themselves awash in Ph. D. graduates who can't find jobs and in 500 applications for any open positions. The space station continues to be funded (in spite of the urgings of 17 scientific societies to the contrary); but the SSC truly is dead and gone.

Yet, we are living in a time of remarkable change. It is the turn of the century, the end of a millenium. Surely the last fifty years have been the golden age of physics. The dramatic technological revolution which is underway is a direct result of the quantum revolution in the first half of this century. But who among even the great physicists of that time--Einstein, Fermi, Pauli, Schroedinger--could have even imagined (or predicted) the characteristics of the world we live in today? In 1950 if you asked a physicist to identify the most important contributions of physics to society, it would surely have been said to be nuclear energy (a physicist was quoted as stating that energy would become so cheap, it could be given away) for the good of mankind, and "the bomb" for its destruction. Yet, fifty years later nuclear energy is stagnant in America, but not the Far East. (Ironically, it can be argued that the bomb saved enormous numbers of lives by preventing conventional wars between the superpowers.) Rather, it is the solid state physics of the electronics industry, the technology surrounding the computer, and lasers which have had and continue to have the greatest impact on modern life. And the computer, oh yes, it would turn us all into impersonal numbers. Instead, it is now the principal manifestation of individuality and independence. So much for conventional wisdom in predicting the future impact of physics on the society.

Solid state physics and particle physics have been underway for about a century. Their progress rests on the results obtained from the more than a trillion dollars spent on physical (distinguished from medical) scientific research, much of it since the current funding structure for such research emerged after WWII and much of it driven by military applications. An outside estimate would be that fifty million dollars has been spent on experiments in nonlinear dynamics and chaos.

Expecting chaos to produce results comparable to these highly developed fields is holding it to a standard which even quantum mechanics could not have met at a similar stage in its development.

Money is the coin of the realm; it determines what experiments can be done and the level of the sophistication of their execution and analysis. Currently there are two programs in experimental chaos research, one in ONR and one in DOE. The total funding is at the level of $3 million. Almost **every** individual particle physics experiment has a yearly budget bigger than the **total yearly** budget for all the experiments in nonlinear science Chaos and nonlinear science will never have a significant impact unless the significant resources needed to conduct advanced experiments are made available.

At the beginning of this century physicists discovered that the laws of classical physics did not apply at small distances (quantum mechanics) or at high velocities (relativity). In the last quarter of this century physicists have found that the laws of classical physics do not hold for (even small numbers of) interacting systems. Qualitatively new kinds of behavior are observed when nonlinearities are introduced into either the coupling between or the dynamics (equations) of classical systems. Collective behaviors of a large number of coupled systems (such as a spatially extended premixed flame) can exhibit dramatically different dynamics than an individual element (a small region of the flame front). Chaotic dynamics and nonlinear science have a domain of applicability which is virtually unexplored--between the molecular world and the macroscopic world--the world of complex interacting elements or systems.

I have no idea whether chaos will become technologically important. I certainly don't want to engage in the kind of hype and speculation, such as the "Holy Grails, Woodstocks of Physics, New Silicon Valleys, or God Particles", which have filled the media over developments in physics in the past ten years. If anything, there seems to be an inverse correlation between media coverage and ultimate technological importance.

My request a simple one, based strictly on an intellectual argument: that chaotic dynamics and nonlinear science be accorded the same respect and opportunities as other groups of ideas and experiments which have emerged over the history of physics. We need time and money to conduct the fundamental experiments on more complicated ideal systems before tangible results will appear in applications in society.

One of the central messages of chaos is the inherent futility in trying to predict the evolution of complex systems. Yet it is precisely such a "prediction" that I challenge the reader to undertake. What is your view of the future of physics? In twenty years what topics in physics will be important? Will we still be trying to develop superconductors or buckyballs for technological devices and, if so, what new principles of physics will be involved? Or is it possible that medical data sent from imbedded microchips and sensors (monitoring bodily characteristics) to computers will be analyzed using algorithms developed from the techniques of chaotic dynamics ((my prediction))--How big a "market" is that?)?

So write down your judgement of what physics will be technologically important in twenty years. Give your assessment of the impact and development of nonlinear science in this time frame. Be sure to include a section on solid state, particle physics and the most important technological applications resulting from the discoveries of basic science in these fields. Put it in an envelope, seal it, mail it to yourself, and then store it somewhere safe. Bring it to the workshop in Mobile that Don Herbert will run in 2015. We'll see who does the best.

Maybe you share the view of a former Fermi Lab physicist, David Lindsay, who recently wrote a book entitled, "The End of Physics". Maybe you share the view of Dick Teresi[16], an award-winning science writer, who, in an op-ed piece in the New York Times, wrote that, in spite of discoveries (sightings) of the top quark, "Unfortunately, like Elvis, physics is still dead". Me, I prefer the words of that great American philosopher, Yogi Berra, "It ain't over till it's over." See ya in 2015.

ACKNOWLEDGEMENTS

I would like to thank Don Herbert for inviting me to this workshop and for giving me the opportunity to present my ideas. I would also like to thank Don, his collegues, and his staff for their hospitality and for their effectiveness at running this workshop. My collaborators, Kay Robbins of the University of Texas at San Antonio and Mohamed el-Hamdi, are responsible in equal measure for this research. Martin Golubitsky and Gemunu Gunaratne are the other two principal participants in research in nonlinear science at the University of Houston. They provide constant counseling on issues of mathematics and theoretical physics, respectively. This research has been funded for 11 years by a grant, NK-00014-K-0613, from the Office of Naval Research.

REFERENCES

1. J. Gleick, *Chaos, The Making of a New Science*; (Penguin, New York, 1986).

2. M. Gorman, P. J. Widman and K. A. Robbins, Phys. Rev. Lett. **52**, 2241-2244 (1984).

3. M. Gorman, P. J. Widman, and K. A. Robbins, Physica D **19**, 255-266 (1986).

4. P. J. Widman, M. Gorman and K. A. Robbins, Physica D **36**, 157-166 (1989).

5. For a review of the theoretical studies of the dynamics of cellular flames see F. A. Williams, *Combustion Theory*, (Benjamin Cummings, Menlo Park, 1985) p. 269-372; A. Bayliss, B. J. Matkowsky, and H. Riecke, in *Numerical Methods for PDE's with Critical Parameters*, edited by H. Kaper and M. Garbey, Kluwer Academic Publishers, 1994; S. B. Margolis and G. I. Sivashinksy, SIAM J. Appl. Math. **44**, 344 (1990); A. Bayliss and B. J. Matkowsky, SIAM J. Appl. Math. **52**, 396 (1992).

6. M. Gorman, M. el-Hamdi and K. A. Robbins, Comb. Sci. and Technol. **98**, 37-44 (1994).

7. M. Gorman, M. el-Hamdi, B. Pearson, and K. A. Robbins, Phys. Rev. Lett., to appear.

8. M. Gorman, C. F. Hamill, M. el-Hamdi and K. A. Robbins, Comb. Sci. and Technol. **98**, 25-36 (1994).

9. M. Gorman, M. el-Hamdi and K. A. Robbins, Comb. Sci. and Technol. **98**, 71-78 (1994)

10. M. Gorman, M. el-Hamdi and K. A. Robbins, Comb. Sci. and Technol. **98**, 79-93 (1994).

11. M. Gorman, M. el-Hamdi and K. A. Robbins, Comb. Sci. and Technol. **98**, 47-56 (1994).

12. M. el-Hamdi, M. Gorman and K. A. Robbins, in *Proceedings of the International Conference on Pulsating Combustion*, Comb. Sci. and Technol. **94**, 87-101 (1993).

13. M. Gorman, "Classification of High Dimensional Chaotic Dynamics Using Power Spectra", in *Second Annual Conference on Nonlinear Dynamical Analysis of the EEG*, edited by B. Jansen and M. Brandt, (World Scientific, 1993) p. 3-12.

14. J. J. Collins and C. J. DeLuca, Phys. Rev. Lett. **73**, 764-767 (1994).

15. J. G. Gollub and H. L. Swinney, Phys. Rev. Lett. **35**, 927-930 (1975).

16. Dick Teresi, New York Times, June 11, 1994, op-ed page.

Michael Gorman is an Associate Professor of Physics at the University of Houston. He received his B.S. degree at Boston College. He earned his Ph.D. (condensed matter physics) at the University of Chicago in 1976. He did four years of postdoctoral research with Harry Swinney, first at the City College of New York and later at the University of Texas at Austin where he studied the transition to turbulence in Taylor-Couette flow. His current research is experimental studies of the chaotic dynamics of fluids. For the past eight years he has been studying the chaotic dynamics of premixed flames. He holds three patents in optics. In 1992 he served as Chair of the Texas Section of the American Physical Society. He is the author of over 40 publications and has had 1.4M in research support over the past eight years.

Applications of Chaos in Biology and Medicine

William L. Ditto

Applied Chaos Laboratory, School of Physics
Georgia Institute of Technology, Atlanta, GA 30332-0430

Abstract. Before its discovery chaos was inevitably confused with randomness and indeterminacy. Because many systems *appeared* random, they were actually thought to *be* random. This was true despite the fact that many of these systems seemed to display intermittent almost periodic behavior before returning to more "random" or irregular motion. Indeed this observation leads to one of the defining features of chaos: the superposition of a very large number of unstable periodic motions. Thus the identification in biological systems of unstable periodic or fixed point behavior consistent with chaos makes new therapeutic strategies possible. Recently we were able to exploit such unstable periodic fixed points to achieve control in two experimental systems: in cardiac tissue and brain tissue.

INTRODUCTION

I will put Chaos into fourteen lines
And keep him there; and let him thence escape
If he be lucky; let him twist and ape
Flood, fire, and demon - his adroit designs
Will strain to nothing in the strict confines
Of this sweet Order, where, in pious rape,
I hold his essence and amorphous shape,
Till he with Order mingles and combines.
Past are the hours, the years, of our duress,
His arrogance, our awful servitude:
I have him. He is nothing more nor less
Than something simple not yet understood;
I shall not even force him to confess;
Or answer. I will only make him good.

Edna St. Vincent Millay

There exists a behavior between rigid regularity and pure chance. Its name is *chaos*. Our scientific paradigms encompass at one extreme the precise clock-like motion of periodic systems and at the other extreme the vagaries of pure chance but have curiously failed to account for the behavior commonly observed between these two extremes -- chaos. When we see irregularity we chauvinistically cling to randomness and disorder as explanations. Why should this be so? Why is it that when we encounter irregularity the whole vast machinery of probability and statistics is belligerently applied? Only recently, however, has the scientific community begun to realize that the tools of chaos theory may be more appropriately applied to the study of irregular systems than probabilistic techniques. We rediscover daily that nature prizes and exploits change and complexity. The existence of chaos has definitively shown that even simple systems can behave in very complex ways. As usual, nature has exhibited a richer, more elegant structure than our limited scientific paradigms expected.

The discovery of chaotic behavior in nature initiated a rapid revolution in the sciences. Chaos has been discovered, accepted and assimilated into the scientific community over the last fifteen years.[1] Yet until recently chaos was viewed as a very mathematical and theoretical discipline. Many were asking "what good is chaos?" and answering that it will never be more than nuisance; something to be avoided in our attempt to develop and use nonchaotic systems. Researchers have demonstrated that chaos admits possibilities and opportunities that simple behavior cannot.

The world around us is dynamic; most things in it, including ourselves, change in time. Consequently, it's not surprising that much of science is dedicated to analyzing and understanding the shape and pace of these changes. We apply our knowledge of dynamics toward the understanding and manipulation of the natural world.

Most man-made systems are designed with simple dynamical behaviors such as stationary states (fixed points) or periodic motion (cycles). Most communications systems rely on the cyclic variation of voltages in circuits (periodic behavior) for sending and receiving messages. This is because simple systems are predictable. Knowing where such simple systems are at one time means we can predict where they will be at times far into the future. Mathematical models of simple systems are usually linear and thus relatively easy with which to work.

Nevertheless, many naturally occurring systems are nonlinear and the outcome of the system's behavior is not simply proportional to the voltages, pressures, or other variables that describe the system. Chaotic behavior frequently occurs in these nonlinear systems and it is *not* predictable very far into the future. Fig. 1 shows a *time series* of position of a driven nonlinear oscillator (as shown, a pendulum) undergoing chaotic motion. Systems evolving chaotically display a *sensitivity to initial conditions*. That is, two nearly identical systems started at slightly different initial states or conditions will soon evolve to values that are vastly different. Such chaotic systems will quickly become completely uncorrelated, even though the overall characteristics of behavior will remain the same.

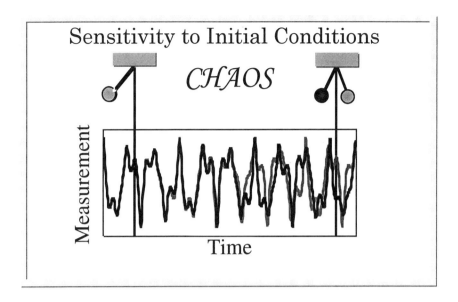

FIGURE 1. Typical time series from a driven system with a nonlinear restoring force (such as the pendula shown) undergoing chaotic motion. Black and grey time series represent pendula started with slightly different initial conditions. This demonstrates the sensitivity to initial conditions and lack of long term prediction that characterize chaotic systems.

To the casual observer chaotic systems appear to behave in a random and noisy fashion. However they are completely deterministic because their irregular motion is generated by the system itself instead of random external influences. Examples of chaos are plentiful in nature and include magnetic systems, population dynamics, chemical reactions, physiology, hydrodynamics and electronic circuits.[2] Thus, what we really want is an understanding of nonlinear dynamics; the way nonlinear systems change in time.

The key to understanding nonlinear systems is to go from their irregular behavior, which can be hard to interpret, to an understanding of the dynamics of the system. A fundamental tool of nonlinear dynamics which accomplishes this is called phase space. The notion is simple. We use the laws of science that are applicable to our system to identify *all* the variables necessary to completely describe our system. Those that can change in time are the dynamical variables and define the phase space. Other variables can be thought of as constant parameters which, once set, never change. We can view the phase space by making a plot in which each axis is associated with a dynamical variable. A point on this plot specifies the complete state of the system at a given time. As the system changes the point traces out a curve which is the system's trajectory in phase space. This trajectory represents the history of the dynamical system. This concept is actually an old one which goes back over 150 years to scientists like William Hamilton and Karl Jacobi who used phase space to gain a geometrical picture of the solutions to equations of motion for physics.

In dynamical systems language the final trajectory of a system is called an *attractor*. Attractors yield an informative geometric view of the motion. In systems with many dynamical variables we would have attractors in a many-dimensional phase space. In biology, we might have a model of the populations of predators and prey such as wolves and rabbits.[3] The dynamical variables will be the population numbers. The relation of the change in these numbers to their present values will be guided by biological principles and experiments. Then the system may settle into a fixed-point attractor, where the wolves eat the rabbits at the same rate that the rabbits multiple. Here the fixed-point attractor would represent a stable unchanging population. Or the system may settle into a periodic attractor, where the wolves starve in alternate years, leaving fewer wolves in off-years, resulting in rabbit overpopulation, wolves feasting, wolf overpopulation, rabbit underpopulation, and back to wolves starving. The phase space diagram would be a looped, periodic attractor which has a period of two years.

Unlike linear systems, chaotic systems have no solutions to the equations which can be written as simple formulas. But we can calculate, in terms of numbers, the form of their attractors. Access to computers has allowed modern scientists finally to view this chaotic motion. It is the geometry of the attractor in phase space shown in Fig. 2 which leads to a fuller understanding of the dynamics. Although the motion of the system eventually settles onto the attractor, small perturbations develop into large differences as shown in Fig. 1. A perturbed system will still be on the attractor at a later time, but at a completely different location from where it would have been without the perturbation. Unlike linear or simple systems, predictability in a chaotic system is difficult even for short times and impossible *in principle* for long times.

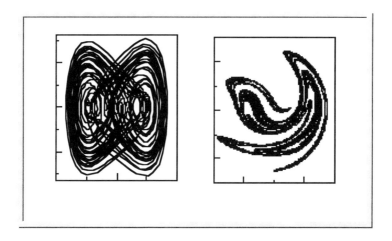

FIGURE 2. Typical phase space portraits of chaotic system. Left: position vs. velocity of system in chaotic regime, Right: position vs. velocity of system strobed at drive frequency (a Poincaré section).

A chaotic attractor cannot have any long-lasting periodicities. That is, motion on it cannot repeat exactly. To see this imagine that we have a trajectory that loops back on itself. Then, unlike linear systems, a slight perturbation will send the chaotic system careening away from its periodic loop. In fact, we can think of an attractor as an infinite collection of unstable periodic orbits with the system's state-space point moving from one to another.

Initially, most "applications" of chaos theory were attempts to simply make dynamical systems avoid chaos entirely – a product of our fixation with simple systems. Lately it is becoming clear that the unique qualities of chaos and other nonlinear dynamical behaviors are desirable for exploitation.

One of the keys to unlocking potential applications of chaos is the collection of unstable periodic orbits composing a chaotic attractor. The infinite number of periodic orbits embedded in chaotic systems represent a vast library of possible behaviors. Thus if we could entice a chaotic system to select a desired available orbit amidst the sea of chaos and to remain near that orbit, the presence of the chaos would prove to be an asset by providing flexibility in moving from one periodic orbit to another.

Such a scheme was recently developed by three theorists at the University of Maryland, Ed Ott, Celso Grebogi, and James Yorke (OGY).[4] Tiring of being asked "what good is chaos?" they blended chaos with control theory through the formulation of a method that forces chaotic systems to stay near selected unstable periodic orbits. The method is conceptually straightforward and appealing. First one obtains information about the chaotic system by constructing a Poincaré section. This is just what it sounds like, a section or slice through the state-space trajectory. One of the simplest ways to do this is shown in Fig. 2 where we have obtained a section from the state space trajectory of the nonlinear driven oscillator by strobing the measurement of the system at the drive frequency. This technique simplifies the analysis of the dynamics because periodic orbits in phase space appear as fixed points in a section. Thus the unstable period-one orbit on the chaotic attractor displayed in Fig. 3 appears as one point. To control a chaotic system one waits until the system comes near a desired unstable fixed point (periodic orbit) in the section and then applies a small perturbation to an available system parameter to keep the motion near the fixed point.

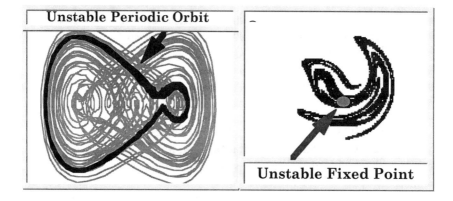

FIGURE 3. Phase space portraits showing an unstable period one orbit (left) and its representation as an unstable period one fixed point (right) in the Poincaré section.

To calculate the control perturbations required to do this one observes from the section how the system approaches the desired fixed point. This observation consists of four parts. First you determine the location of the unstable fixed point, second you determine the directions along which the motion converges and diverges from this unstable fixed point. These directions are known as stable and unstable directions (eigenvectors) respectively as shown in Fig. 4. Third the rates of convergence and divergence (stable and unstable manifolds) must be obtained. Lastly, the movement of the stable and unstable directions with alterations of the control parameter must be determined. These four observations will allow us to continually nudge the system back onto the unstable periodic orbit.

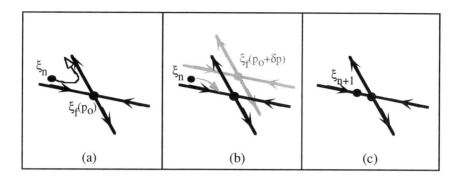

FIGURE 4. Schematic of OGY control algorithm for a saddle fixed point: (a) the nth iterate ξ_n falls near the fixed point $\xi_f(p_0)$. (b) Turn on the perturbation of p to move the fixed point. (c) The next iterate is forced onto the stable manifold of $\xi_f(p_0)$. Turn off the perturbation. Note that ustable manifold has outward pointing arrows and the stable manifold has inward pointing arrows.

While the OGY control method does not require prior knowledge of the system and no model for the system is required, several other requirements must be met. You must be able to construct a Poincaré section of the motion (although this requirement can be relaxed using variations of the method[5]). The system parameter chosen for the control must be capable of moving the chaotic attractor. Additionally, you must be able to change the control parameter quickly compared to the response time of the system. The control perturbations are computed by determining how far the system is from the desired fixed point. The control signals are then applied in real time to nudge the motion onto the stable direction of the fixed point and hence onto the fixed point. This stabilizes, for the moment, the desired unstable periodic motion. The OGY method assumes only the following four points:

(1) The dynamics of the system can be represented as arising from an n-dimensional nonlinear map (*e.g.*, by a surface of section or time one return map), the iterates being given by $\vec{\xi}_{n+1} = \vec{f}(\vec{\xi}_n, p)$ where p is some accessible system parameter.

(2) There is a specific periodic orbit of the map which lies in the attractor and around which one wishes to stabilize the dynamics.

(3) There is a maximum perturbation $\delta p*$ in the parameter p, by which it is acceptable to vary p from the nominal value p_0.

(4) The position of the periodic orbit is a function of p, but the local dynamics about it do not vary much with the allowed small changes in p.

Note that while the dynamics is assumed to arise from a map, one needs no model for the global dynamics. These assumptions would seem to allow for the control of any chaotic system for which a faithful Poincaré section can be constructed. The construction of a map from and the location of periodic orbits in experimental data is a straightforward process.

To control chaotic dynamics one only needs to learn the *local* dynamics around the chosen unstable periodic orbit by observing iterates of the map near the desired orbit and fitting them to a local linear approximation of the map. From this, one can find the stable and unstable eigenvalues as well as the local stable and unstable manifolds (given by the eigenvectors). Next, by changing p slightly and observing how the desired orbit changes position, one can estimate the change of the orbit location with respect to p.

To control the chaos, one attempts to confine the iterates of the map to a small neighborhood of the desired orbit. When an iterate falls near the desired orbit, we change p from its nominal value p_0 by δp, thereby changing the location of the orbit and its stable manifold, such that the *next* iterate will be forced back toward the stable manifold of the *original* orbit (Fig. 4 illustrates this method for the case of a saddle fixed point located at $\vec{\xi}_f(p_0)$).

The OGY control method is strongly analogous to a marble rolling down the middle of a saddle. As one rolls the marble repeatedly down the saddle it falls off one side or another. The center of the saddle represents the desired unstable periodic orbit. The saddle has both a stable and unstable direction. The marble is forced to linger indefinitely at the center of the saddle by moving the saddle to make the marble continually fall towards the center. Thus the stable and unstable directions, if moved, quickly roll the marble onto the desired point. In chaotic systems the slopes of the saddle can be quite steep and, consequently, through the use of sensitivity to initial conditions, we can make the system do our bidding with the application of *very small* control signals.

This type of control, known to control theorists as proportional feedback, is certainly not new. What is new, however, is the realization that the presence of chaos can greatly enhance the effectiveness of the control. This counter-intuitive result arises from the defining feature of chaotic systems — *sensitivity to initial conditions*. This provides the chaotic system with tremendous *sensitivity to implemented controls.*

No theory ever survives intact in contact with experiment. The OGY theory, however, was to provide the exception that proves the rule. Mark Spano, Steve Rauseo and I were able to implement the first experimental control of a chaotic system.[6] This experiment, which has been used to demonstrate a tremendous variety of chaotic behavior, is built around a magnetoelastic ribbon whose stiffness can be varied (by up to a factor of 17!) through the application of small magnetic fields. The ribbon, which resembles a short stiff strand of Christmas tree tinsel, was clamped at the base and oriented vertically as shown in Fig. 5. The position of the ribbon is measured once each period, $T \equiv 1/f \approx 1Hz$, of the driving field by means of a beam of light directed at a small spot on the ribbon. In practice it is not always possible to measure all of the variables that comprise the phase space of a dynamical system. If you are limited to one measurement as a function of time, it is possible to reconstruct the phase space of your system by using a *delay coordinate embedding*. In this method, the sensor's output $X_i = X(i\Delta)$, where Δ is the sampling time, are arranged into delay coordinate vectors whose d components at time $i\Delta$ are $(X_i, X_{i+n}, X_{i+2n}, ..., X_{i+(d-1)n})$. As described above we found the stable and unstable eigenvectors, as well as the change in the fixed point position with respect to a change in the system parameter we chose for control - the dc magnetic field, H_{dc}. (In theory, either of the three system parameters, H_{dc}, H_{ac} or f, could be used to control the chaos.) The Poincaré section was easily obtained by triggering the voltmeter directly from the frequency synthesizer. Consequently, our phase space consists of vectors of the form (X_i, X_{i+1}) which define a two dimensional map. The sampling time Δ is the period T of the ac magnetic drive field. The computer measuring this phase space in real time also controls the strengths and frequencies of the magnetic fields.

We wait for the system to visit a region of phase space that is within some radius, ε, of the target fixed point. Then it is only necessary to form the delay coordinate vector, calculate the change in the position of the fixed point that is needed (Fig. 6) and finally find the corresponding adjustment to the dc magnetic field that is required to achieve that change.

FIGURE 5. Magnetoelastic ribbon experiment.

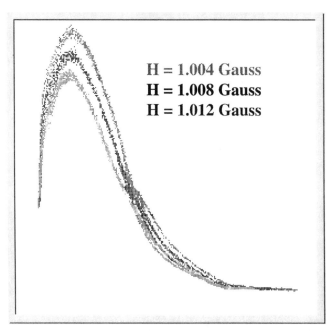

FIGURE 6. Poincaré sections constructed from ribbon experimental data for different values of the control parameter H_{dc}.

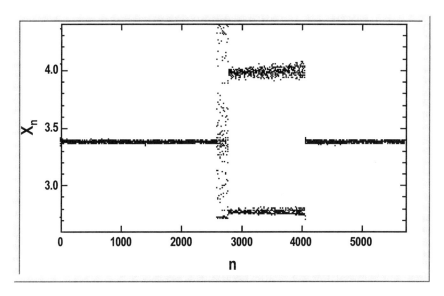

Figure 7. Position versus data point number demonstrating controlled period one, chaos, controlled period two, chaos and finally controlled period one again.

CONTROL OF CHAOS IN HEART TISSUE

We have found that it is possible to control a chaotic cardiac arrhythmia using the same basic properties of chaotic systems that were exploited by OGY but which are here employed in a new method of chaos control suitable for use in systems where no systemwide parameters can be readily manipulated, as is required by the OGY method.[7]

Our cardiac preparation consisted of an isolated perfused portion of the interventricular septum from a rabbit heart as shown in Fig. 8. The heart was stimulated by passing a 3 ms constant voltage pulse, typically 10-30 volts, at twice threshold between platinum electrodes embedded in the preparation. Electrical activity was monitored by recording monophasic action potentials with Ag-AgCl wires on the surface of the heart. Monophasic action potentials were digitized at 2 kHz and processed in real time by a computer to detect the activation time of each beat from the maximum of the first derivative of the voltage signal.

Figure 8. Rabbit heart experiment.

Arrhythmias were induced by adding 2 to 5 µM ouabain with or without 2 to 10 µM epinephrine to the arterial perfusate. The mechanism of ouabain/epinephrine-induced arrhythmias is probably a combination of triggered activity and non-triggered automaticity caused by progressive intracellular Ca^{2+} overload from Na^{+} pump inhibition and increased Ca^{2+} current. Typically the ouabain/epinephrine combination induced spontaneous beating, initially at a constant interbeat interval and then progressing to period 2 and higher order periodicity before developing a highly irregular aperiodic pattern of spontaneous activity. The duration of the aperiodic phase was variable, lasting up to several minutes before spontaneous electrical activity irreversibly ceased. The spontaneous activity induced by ouabain/epinephrine in this preparation showed a number of features symptomatic of chaos. Most importantly, in progressing from spontaneous beating at a fixed interbeat interval to highly aperiodic behavior, the arrhythmia passed through a series of transient stages that involved higher order periodicities. These features are illustrated in Figs. 9-11 in which the nth interbeat interval (I_n) has been plotted against the previous interval (I_{n-1}) at various stages during ouabain/epinephrine-induced arrhythmias. Such a *Poincaré return map* allows us to view the dynamics of the system as a sequence of pairs of points (I_n, I_{n-1}), thus converting the dynamics of our system to a map.

Period One

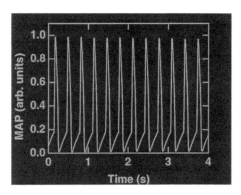
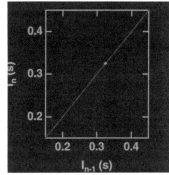

Period Two

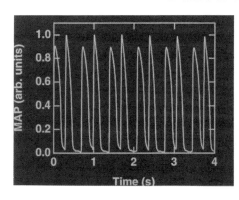
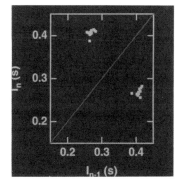

Figure 9. Periodic waveforms (left) and return maps interbeat intervals (right) for the rabbit heart experiment.

Period Four

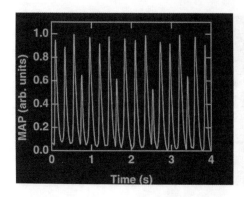
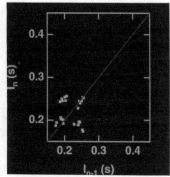

Period Five

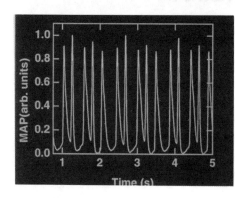
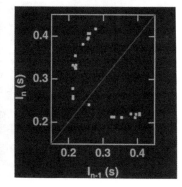

Figure 10. More periodic waveforms (left) and return maps of interbeat intervals (right) for the rabbit heart experiment.

Cardiac Chaos

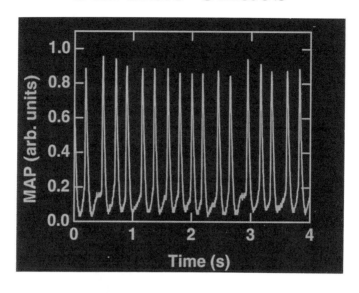

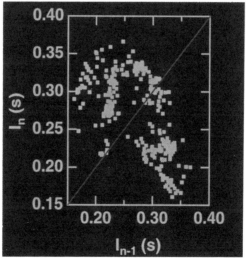

Figure 11. Chaotic waveforms (top) and return maps of interbeat intervals (bottom) for the rabbit heart experiment.

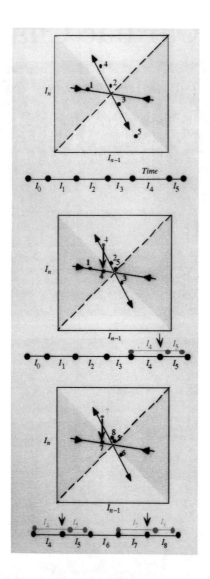

FIGURE 12. Control of chaos in excitable tissue. (top) a return plot of I_n vs. I_{n-1} typically constructed from interval data. Note the presence of an unstable fixed point along the 45 degree line. (middle) demonstrates how to pull the interval down onto the stable direction (manifold) and consequently onto the unstable fixed point. The interbeat intervals and the control stimuli are shown below the return map. (The grayed intervals are without control.) (bottom) depicts the subsequent behavior of the system under control. Note the application of stimuli (arrows) is only intermittent.

In this preparation our method of chaos control consisted of delivering a perturbation (near the desired fixed point) which forces the system state point onto the stable manifold of the desired fixed point as shown in Fig 12.

In terms of the saddle analogy, we can no longer move the saddle to stabilize the marble, so we poke the marble to keep it on the saddle. We developed this method when it became obvious that our cardiac preparation possessed no systemwide parameter that could be changed sufficiently quickly to implement classical OGY control.

Chaos control with this approach was complicated by the fact that in this experiment intervention was, of necessity, unidirectional By delivering an electrical stimulus before the next spontaneous beat the interbeat interval could be shortened but it could not directly be lengthened. This is because a stimulus which elicits a beat from the heart must, by definition, shorten the interbeat interval between the previous spontaneous beat and the next expected spontaneous beat.

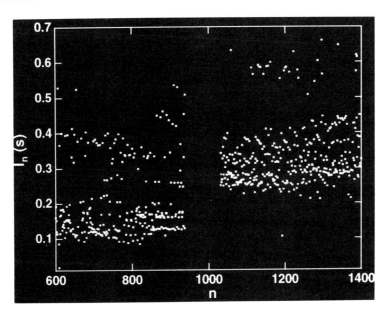

FIGURE 13. Chaos control in heart tissue. The region of chaos control (dark grey) is preceded by a brief learning phase when the parameters necessary for control are discerned from the data.

Several observations should be made about the pattern of the stimuli delivered by the chaos control program. First, these stimuli did not simply overdrive the heart. Stimuli were delivered sporadically, not on every beat and approximately once in every three beats on average. In contrast, periodic pacing, in which stimuli were delivered at a fixed rate, was never effective at restoring a periodic rhythm and often made the aperiodicity more marked.

CONTROL OF CHAOS IN BRAIN TISSUE

Our success in controlling chaos in the rabbit heart tissue preparation led us to see if a similar strategy could control chaotic behavior in brain tissue.[8] One of the hallmarks of the human epileptic brain is the presence of brief bursts of focal neuronal activity known as interictal spikes. Often such spikes trigger epileptic seizures in a nearby region of the brain. Several types of *in vitro* brain slice preparations, usually after exposure to convulsant drugs that reduce neuronal inhibition, exhibit population burst-firing activity similar to the interictal spike. One of these preparations is the high potassium [K^+] model, where slices from the hippocampus of the temporal lobe of the rat brain (a frequent site of epileptogenesis in the human) are exposed to artificial cerebrospinal fluid containing high [K^+] which causes spontaneous bursts of synchronized neuronal activity which originate in a region known as the third part of the *cornu ammonis* or CA3, as shown in Fig. 14.

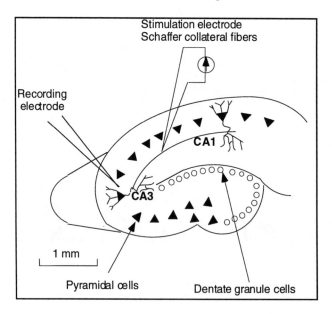

Fig 14. Schematic diagram of the transverse hippocampal brain slice and arrangement of recording electrodes.

If one observes the timing of these bursts, clear evidence for unstable fixed points is seen in the return map. As reported, we were able to regularize the timing of such bursts through intervention with stimuli delivered with timing as dictated by chaos control in order to put the system onto the stable direction (manifold). As shown in Fig. 15, not only were we able to regularize the intervals between spikes but we were also able to make the intervals more chaotic through a chaos maintenance control strategy . It is the latter which might serve a useful purpose in breaking up seizure activity through the prevention or eradication of pathological order in the timing of the spikes.

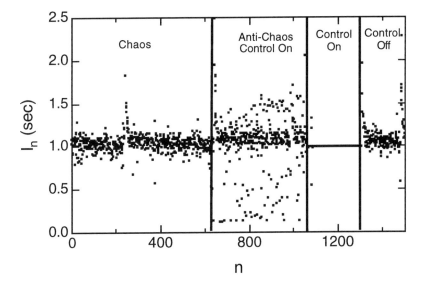

FIGURE 15. Demonstration of maintenance of chaos and of chaos control in a hippocampal slice of a rat brain exposed to artificial cerebrospinal fluid containing high [K^+] and undergoing spontaneous chaotic population burst firing or spiking. The interspike interval I_n (sec) is plotted versus spike number n.

MAINTENANCE OF CHAOS

Since Ott, Grebogi and Yorke published their paper on the theory of the control of chaos, a major thrust of the work in experimental chaos has been to convert the chaos found in various physical systems into periodic motion. Once the experimental control of chaos was first demonstrated in a mechanical system (a magnetoelastic ribbon), the control of chaos was subsequently implemented in lasers, electronic circuits, chemical reactions and biological systems. Although chaos control may be very advantageous in many systems, it has been suggested that pathological destruction of chaotic behavior (possibly due to some underlying disease) may be implicated in heart failure and some types of brain seizures.[9] Thus some systems may require chaos and/or complexity in order to function properly. Another situation in which the maintenance of chaos might be useful is the mixing of fluids. Experimental work by Schiff and colleagues[8] demonstrated an *ad hoc* method for increasing the complexity (decreasing the periodicity) of an *in vitro* hippocampal rat brain slice preparation. Recent theoretical and computational work by Yang et al.[10] indicates that intermittent chaotic systems can be made to exhibit continuous chaotic behavior (no intermittent periodic episodes).

We have formulated a general theoretical method for the maintenance of chaos, which was then implemented experimentally in the ribbon experiment which was configured to exhibit intermittent chaos. This intermittency appears as chaos interspersed with long periodic episodes. This method is readily applicable to experiment and relies only on experimentally measured quantities for its implementation.

Our anticontrol method[11] makes only the following assumptions about the system: (1) the dynamics of the system can be represented as an n-dimensional nonlinear map (*e.g.*, by a surface of section or a return map) such that points or iterates on such a map are given by $\vec{\xi}_n = \vec{f}(\vec{\xi}_{n-1}, p)$, where p is some accessible system parameter; (2) there is at least one specific region of the map (termed a "loss region") that lies on the attractor into which the iterates will fall when making the transition from chaos to periodicity; and (3) the structure of the map does not change significantly with small changes $\delta p \equiv p - p_0$ in the control parameter p about some initial value p_0.

On the return map derived from the ribbon, the locations of loss regions are determined by observing immediate preiterates of undesired fixed points which correspond to periodic orbits. Clusters of these preiterates are identified as the loss regions. The extent of each loss region is determined by the distribution of points in that region. The time evolution of each region may be traced back through m preiterates, as desired.

Next, in a fashion similar to the OGY chaos control method, we change p slightly, observe the resulting change in each loss region's location and estimate the local shift of the attractor, $\bar{\mathbf{g}}$, for each loss region with respect to a change in p as follows:

$$\bar{\mathbf{g}} = \frac{\partial \vec{f}(\vec{\xi}_n, p)}{\partial p} \approx \frac{\Delta \vec{f}(\vec{\xi}_n, p)}{\Delta p} \qquad (1)$$

As an approximation we take this shift to be a constant value for all loss regions on the attractor for sufficiently small parameter changes δp (otherwise calculation of $\bar{\mathbf{g}}$ for each loss region is required). This is not strictly necessary in order to implement the method but is simply a convenience that is approximately true for many systems (including our magnetoelastic ribbon) and for small δp's.

Anticontrol can be applied once the system has entered the mth preiterate of the loss region. Since the map is constructed as a return map (or delay coordinate embedding) with ξ_n versus ξ_{n-1}, the y-coordinate of the nth point becomes the x-coordinate of the $(n+1)^{st}$ point. Because we know the x-coordinate of the next point and the *size* of the region that this $(n+1)^{st}$ point would normally fall into, we calculate a minimum distance that we must move the attractor so that this next point falls outside of that region. This distance d is translated into the appropriate parameter change δp by

$$\delta p_n = \frac{d_{n+1}}{|\bar{\mathbf{g}}|} \tag{2}$$

where the direction of the motion is along $\bar{\mathbf{g}}$.

If each of the m preiterates of the loss region is circumscribed by a circle of radius r_m (the worst case), we have $\delta p_n = 2 r_m / |\bar{\mathbf{g}}|$, where it is understood that the $(n+1)^{st}$ point falls into the mth preiterate region. This is the maximum perturbation needed to achieve anticontrol and guarantees that the next point will fall outside the mth preiterate region by moving the point one full diameter of the circle surrounding the loss region. We can improve upon this worst case. With a return map, we know the x-coordinate of the next point. Because we have the choice of whether to apply the perturbation in either the positive or the negative $\bar{\mathbf{g}}$ direction, we can select the sign of the perturbation to move the next point to the left if this x-coordinate is in the left half of the preiterate region and vice versa. Thus the minimum distance to move is reduced to r_m and consequently $\delta p_n = r_m / |\bar{\mathbf{g}}|$, a significant reduction in the strength of the perturbation.

Additionally, if the shape of the preiterate region of interest is approximately linear (line-like) and its slope is perpendicular to $\bar{\mathbf{g}}$, then d is at most r_m and may approach the thickness of this linear segment ($\delta p_n << r_m / |\bar{\mathbf{g}}|$). Thus, while not necessary to achieve anticontrol, a detailed knowledge of the *shape* of the loss region and its preiterates can further reduce the size of the perturbation required to achieve anticontrol.

To experimentally implement the anticontrol algorithm, we measure the position ξ_n of a point on the ribbon once every driving period. We then construct a return map (delay coordinate embedding) by plotting the current position ξ_n versus ξ_{n-d}, where d is the delay. Here, where the data is strobed at the driving frequency, we have $d = 1$.

From the return map, we have identified the loss region and its preiterates (circles in Fig. 16). The loss region (R_L) is denoted by the circle immediately to the left of the diagonal. Its first preiterate (R_{L-1}) lies to the right of the diagonal. The other circles denote earlier preiterates (R_{L-m}). A point that enters any of these preiterate regions will eventually go to the region R_{L-1}. Once there the point will proceed into the loss region R_L on the next iterate. Then the system goes into periodic motion. This appears as a cluster of points (R_P) on the diagonal of the map. The points that enter the preiterate regions mediate the intermittent transition from chaos to periodicity. During anticontrol we apply a perturbation when an orbit enters the region R_{L-1} so that the next orbit will fall out of the region R_L.

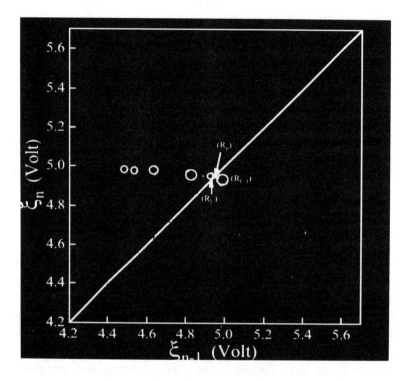

Figure 16. The intermittently chaotic attractor with the approach to the loss region that leads to capture indicated by the circles.

The extent of the mth preiterate region is determined by observing the set of points that after m iterations fall into the loss region as well as neighboring points that do not fall into the loss region after m iterations. The boundary of the loss region lies between these points. The $\bar{\bar{g}}$ vector is determined by changing p_0.

During anticontrol the largest perturbation is 1.106% of the nominal dc magnetic field. Significantly the anticontrol signal needed to be applied *only 0.12%* of the entire anticontrol time to keep the system chaotic.

The effect of the anticontrol may be qualitatively appreciated by looking at Figs. 17 and 18. The 3-dimensional histogram in Fig. 17 reflects the density of points over the attractor (return map) of the unperturbed system. (It is important to note that the vertical scale here is *linear*.) Observe that most of

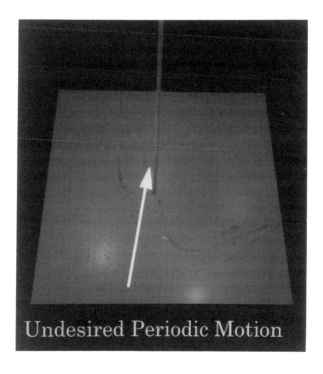

Figure 17. 3-dimensional histogram of the unperturbed data distributed over the attractor (return map). The scale is enhanced by a factor of 10 (relative to Fig. 18) in order to make visible the chaotic transients present in the data. The period 1 peak (which lies on the diagonal line) is so strong that it would be off the scale even had the vertical scale not been enhanced by this factor of 10.

the probability resides in the strong central peak that represents the period 1 orbit. The density of the points resulting from anticontrol is presented in Fig. 18. Here the probability is spread over the entire chaotic part of the attractor with a distribution that approximates that of the chaotic parts of the unperturbed, intermittently chaotic system.

Note that there is still a period 1 peak in Fig. 18. This represents a period 1 motion that *does not destroy the chaos*. The anticontrol only prevents the period 1 motion initiated by following the sequence of preiterates that leads to the loss region and subsequently to the loss of chaoticity. However the method does *not* interfere with sequences of points that enter the loss region by other routes and that do not destroy the chaos. This period 1 motion is *naturally* unstable and is properly one of the unstable periodic motions that comprise the chaos itself. Hence it is not removed.

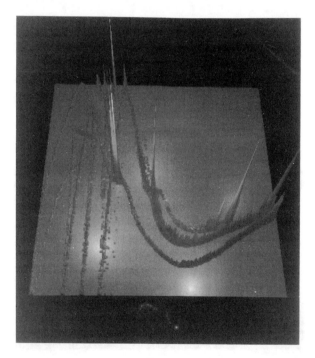

Figure 18. 3-dimensional histogram of the anticontrolled data distributed over the attractor.

To reiterate, we interrupt only the sequence of preiterates that lead to entrapment in a periodic motion. It is because we take this approach, rather than the approach of completely excluding the system from the region of phase space around the unstable periodic motion, that we are able to maintain the chaos with only rare interventions (~0.12% of the time).

It is anticipated that the anticontrol of chaos will have bearing on a wide range of physical and biological systems where chaos and complexity are found to be desirable. One application that concerns the authors is to the control of epileptic seizures. Of course, it may also be of use in other biological systems such as the heart, which may need chaos and complexity in order to function properly.

CONCLUDING REMARKS

From the practical standpoint several difficulties still exist for the control of chaos to make the transition from the lab into applications. One of these includes the difficulties associated with acquisition of desired unstable periodic orbits. If a periodic orbit chosen for control is rarely visited by the uncontrolled system then an unacceptable amount of time may elapse before the system comes near enough to the orbit to attempt control. Troy Shinbrot and colleagues at the University of Maryland and the Naval Surface Warfare Center have demonstrated, both theoretically and experimentally (on the magnetic ribbon experiment)[12] a method for the rapid targeting and acquisition of unstable periodic orbits in chaotic systems. They exploit the sensitivity of the chaotic system to small perturbations to rapidly move the uncontrolled system from an arbitrary initial state to an arbitrary accessible desired state. This method, termed *targeting*, requires that a map corresponding to the experimental Poincaré section be extracted by observing the system during a learning period. The learned map is then used to perturb the control parameter such that the system is taken from the current state to the desired state in a much shorter time than allowing the system to enter the state naturally. Through the use of targeting in the magnetoelastic ribbon experiment, Troy and colleagues were able to reduce the time required to acquire rarely visited unstable orbits by factors as high as 25. Thus the method of targeting provides another example where the presence of chaos can be exploited to improve system performance.

As we have seen, the ability to exploit chaos now makes it advantageous to discard linearity, one of the classical dogmas of the engineer and to design devices using, rather than avoiding, nonlinearity and chaos. Through the use of chaos we may be able to replace many linear systems that do only one thing well with more flexible nonlinear systems. In this way we someday may be able to make physical devices that mimic the flexibility of biological systems.

ACKNOWLEDGMENTS

I would like to thank ONR for their continued support of the Applied Chaos Lab. Additionally I would like to thank my wife Robin, without whom none of this would have been possible.

REFERENCES

1. James Gleick, *Chaos* (Viking, New York, 1987).
2. W. L. Ditto and L. M. Pecora, *Scientific American* **269**, 78 (1993).
3. R. May, *Science* **186**, 645 (1974).
4. E. Ott, C. Grebogi and J. A. Yorke, *Phys. Rev. Lett.* **64**, 1196 (1190).
5. T. A. Shinbrot, C. Grebogi, E. Ott, and J. A. Yorke, *Nature* **363**, 411 (1993).
6. W. L. Ditto, S. N. Rauseo and M. L. Spano, *Phys. Rev. Lett.* **65**, 3211 (1990).
7. A. Garfinkel, M. L. Spano, W. L. Ditto and J. N. Weiss, *Science* **257**, 1230 (1992).
8. S. J. Schiff., K. Jerger, D. H. Duong, T. Chang, M. L. Spano, and W. L. Ditto, *Nature* **370**, 615 (1994).
9. A. L. Goldberger, D. R. Rigney and B. J. West, *Scientific American* **262**, 42 (1990).
10. W. Yang, M. Ding, A. Mandell and E. Ott, *Phys. Rev. E.* **51**, 102 (1995).
11. V. In, S. E. Mahan, W. L. Ditto and M. L. Spano, *Phys. Rev. Lett.* **74**, 4420 (1995).
12. T. A. Shinbrot, C. Grebogi, E. Ott, and J. A. Yorke, *Nature* **363**, 411 (1993).

William Ditto graduated from the University of California at Los Angeles with a B.S. degree in Physics in 1980. In 1988 he received his Ph.D. in Theoretical Physics (Field Theory) from Clemson University. Dr. Ditto subsequently worked for the Department of the Navy in Washington, DC where he demonstrated the first experimental control of chaos. As a consequence, his work has twice been featured on the cover of *Science News*, and has been reported in numerous articles in *Science*, *Nature*, *Scientific American*, and *Time* magazine. His work has led to practical applications of chaos research. He has licensed control of chaos technology to the health care industry with the goal of developing smart heart defibrillators to reduce fatalities in heart attack victims. He is also working on control of chaos strategies for the possible control of epilepsy. Dr. Ditto is currently developing an applied chaos laboratory in the school of physics at the Georgia Institute of Technology where he is a Professor in the Department of Physics. He has written numerous articles on chaos including an article in the August 1993 issue of *Scientific American*. He has organized numerous conferences on the applications of chaos and is currently writing a book for Cambridge University Press on practical applications of chaos entitled *Applied Chaos*.

AUTHOR INDEX

B

Bassingthwaighte, J. B., 54
Beard, D. A., 54

C

Campbell, D. K., 115

D

Ditto, W. L., 175

G

Gilmore, R., 35
Gorman, M., 158

H

Hayles, N. K., 133
Herbert, D. E., 1

P

Percival, D. B., 54

R

Raymond, G. M., 54

S

Stewart, H. B., 80

AIP Conference Proceedings

	Title	L.C. Number	ISBN
No. 226	The Living Cell in Four Dimensions (Gif sur Yvette, France 1990)	91-55209	0-88318-794-9
No. 227	Advanced Processing and Characterization Technologies (Clearwater, FL 1991)	91-55194	0-88318-910-0
No. 228	Anomalous Nuclear Effects in Deuterium/ Solid Systems (Provo, UT 1990)	91-55245	0-88318-833-3
No. 229	Accelerator Instrumentation (Batavia, IL 1990)	91-55347	0-88318-832-1
No. 230	Nonlinear Dynamics and Particle Acceleration (Tsukuba, Japan 1990)	91-55348	0-88318-824-4
No. 231	Boron-Rich Solids (Albuquerque, NM 1990)	91-53024	0-88318-793-4
No. 232	Gamma-Ray Line Astrophysics (Paris-Saclay, France 1990)	91-55492	0-88318-875-9
No. 233	Atomic Physics 12 (Ann Arbor, MI 1990)	91-55595	088318-811-2
No. 234	Amorphous Silicon Materials and Solar Cells (Denver, CO 1991)	91-55575	088318-831-7
No. 235	Physics and Chemistry of MCT and Novel IR Detector Materials (San Francisco, CA 1990)	91-55493	0-88318-931-3
No. 236	Vacuum Design of Synchrotron Light Sources (Argonne, IL 1990)	91-55527	0-88318-873-2
No. 237	Kent M. Terwilliger Memorial Symposium (Ann Arbor, MI 1989)	91-55576	0-88318-788-4
No. 238	Capture Gamma-Ray Spectroscopy (Pacific Grove, CA 1990)	91-57923	0-88318-830-9
No. 239	Advances in Biomolecular Simulations (Obernai, France 1991)	91-58106	0-88318-940-2
No. 240	Joint Soviet-American Workshop on the Physics of Semiconductor Lasers (Leningrad, USSR 1991)	91-58537	0-88318-936-4
No. 241	Scanned Probe Microscopy (Santa Barbara, CA 1991)	91-76758	0-88318-816-3
No. 242	Strong, Weak, and Electromagnetic Interactions in Nuclei, Atoms, and Astrophysics: A Workshop in Honor of Stewart D. Bloom's Retirement (Livermore, CA 1991)	91-76876	0-88318-943-7
No. 243	Intersections Between Particle and Nuclear Physics (Tucson, AZ 1991)	91-77580	0-88318-950-X

	Title	L.C. Number	ISBN
No. 244	Radio Frequency Power in Plasmas (Charleston, SC 1991)	91-77853	0-88318-937-2
No. 245	Basic Space Science (Bangalore, India 1991)	91-78379	0-88318-951-8
No. 246	Space Nuclear Power Systems (Albuquerque, NM 1992)	91-58793	1-56396-027-3 1-56396-026-5 (pbk.)
No. 247	Global Warming: Physics and Facts (Washington, DC 1991)	91-78423	0-88318-932-1
No. 248	Computer-Aided Statistical Physics (Taipei, Taiwan 1991)	91-78378	0-88318-942-9
No. 249	The Physics of Particle Accelerators (Upton, NY 1989, 1990)	92-52843	0-88318-789-2
No. 250	Towards a Unified Picture of Nuclear Dynamics (Nikko, Japan 1991)	92-70143	0-88318-951-8
No. 251	Superconductivity and its Applications (Buffalo, NY 1991)	92-52726	1-56396-016-8
No. 252	Accelerator Instrumentation (Newport News, VA 1991)	92-70356	0-88318-934-8
No. 253	High-Brightness Beams for Advanced Accelerator Applications (College Park, MD 1991)	92-52705	0-88318-947-X
No. 254	Testing the AGN Paradigm (College Park, MD 1991)	92-52780	1-56396-009-5
No. 255	Advanced Beam Dynamics Workshop on Effects of Errors in Accelerators, Their Diagnosis and Corrections (Corpus Christi, TX 1991)	92-52842	1-56396-006-0
No. 256	Slow Dynamics in Condensed Matter (Fukuoka, Japan 1991)	92-53120	0-88318-938-0
No. 257	Atomic Processes in Plasmas (Portland, ME 1991)	91-08105	0-88318-939-9
No. 258	Synchrotron Radiation and Dynamic Phenomena (Grenoble, France 1991)	92-53790	1-56396-008-7
No. 259	Future Directions in Nuclear Physics with 4π Gamma Detection Systems of the New Generation (Strasbourg, France 1991)	92-53222	0-88318-952-6
No. 260	Computational Quantum Physics (Nashville, TN 1991)	92-71777	0-88318-933-X
No. 261	Rare and Exclusive B&K Decays and Novel Flavor Factories (Santa Monica, CA 1991)	92-71873	1-56396-055-9

Title	L.C. Number	ISBN
No. 262 Molecular Electronics—Science and Technology (St. Thomas, Virgin Islands 1991)	92-72210	1-56396-041-9
No. 263 Stress-Induced Phenomena in Metallization: First International Workshop (Ithaca, NY 1991)	92-72292	1-56396-082-6
No. 264 Particle Acceleration in Cosmic Plasmas (Newark, DE 1991)	92-73316	0-88318-948-8
No. 265 Gamma-Ray Bursts (Huntsville, AL 1991)	92-73456	1-56396-018-4
No. 266 Group Theory in Physics (Cocoyoc, Morelos, Mexico 1991)	92-73457	1-56396-101-6
No. 267 Electromechanical Coupling of the Solar Atmosphere (Capri, Italy 1991)	92-82717	1-56396-110-5
No. 268 Photovoltaic Advanced Research & Development Project (Denver, CO 1992)	92-74159	1-56396-056-7
No. 269 CEBAF 1992 Summer Workshop (Newport News, VA 1992)	92-75403	1-56396-067-2
No. 270 Time Reversal—The Arthur Rich Memorial Symposium (Ann Arbor, MI 1991)	92-83852	1-56396-105-9
No. 271 Tenth Symposium Space Nuclear Power and Propulsion (Vols. I–III) (Albuquerque, NM 1993)	92-75162	1-56396-137-7 (set)
No. 272 Proceedings of the XXVI International Conference on High Energy Physics (Vols. I and II) (Dallas, TX 1992)	93-70412	1-56396-127-X (set)
No. 273 Superconductivity and Its Applications (Buffalo, NY 1992)	93-70502	1-56396-189-X
No. 274 VIth International Conference on the Physics of Highly Charged Ions (Manhattan, KS 1992)	93-70577	1-56396-102-4
No. 275 Atomic Physics 13 (Munich, Germany 1992)	93-70826	1-56396-057-5
No. 276 Very High Energy Cosmic-Ray Interactions: VIIth International Symposium (Ann Arbor, MI 1992)	93-71342	1-56396-038-9
No. 277 The World at Risk: Natural Hazards and Climate Change (Cambridge, MA 1992)	93-71333	1-56396-066-4
No. 278 Back to the Galaxy (College Park, MD 1992)	93-71543	1-56396-227-6

	Title	L.C. Number	ISBN
No. 279	Advanced Accelerator Concepts (Port Jefferson, NY 1992)	93-71773	1-56396-191-1
No. 280	Compton Gamma-Ray Observatory (St. Louis, MO 1992)	93-71830	1-56396-104-0
No. 281	Accelerator Instrumentation Fourth Annual Workshop (Berkeley, CA 1992)	93-072110	1-56396-190-3
No. 282	Quantum 1/f Noise & Other Low Frequency Fluctuations in Electronic Devices (St. Louis, MO 1992)	93-072366	1-56396-252-7
No. 283	Earth and Space Science Information Systems (Pasadena, CA 1992)	93-072360	1-56396-094-X
No. 284	US-Japan Workshop on Ion Temperature Gradient-Driven Turbulent Transport (Austin, TX 1993)	93-72460	1-56396-221-7
No. 285	Noise in Physical Systems and 1/f Fluctuations (St. Louis, MO 1993)	93-72575	1-56396-270-5
No. 286	Ordering Disorder: Prospect and Retrospect in Condensed Matter Physics: Proceedings of the Indo-U.S. Workshop (Hyderabad, India 1993)	93-072549	1-56396-255-1
No. 287	Production and Neutralization of Negative Ions and Beams: Sixth International Symposium (Upton, NY 1992)	93-72821	1-56396-103-2
No. 288	Laser Ablation: Mechanismas and Applications-II: Second International Conference (Knoxville, TN 1993)	93-73040	1-56396-226-8
No. 289	Radio Frequency Power in Plasmas: Tenth Topical Conference (Boston, MA 1993)	93-72964	1-56396-264-0
No. 290	Laser Spectroscopy: XIth International Conference (Hot Springs, VA 1993)	93-73050	1-56396-262-4
No. 291	Prairie View Summer Science Academy (Prairie View, TX 1992)	93-73081	1-56396-133-4
No. 292	Stability of Particle Motion in Storage Rings (Upton, NY 1992)	93-73534	1-56396-225-X
No. 293	Polarized Ion Sources and Polarized Gas Targets (Madison, WI 1993)	93-74102	1-56396-220-9
No. 294	High-Energy Solar Phenomena: A New Era of Spacecraft Measurements (Waterville Valley, NH 1993)	93-74147	1-56396-291-8
No. 295	The Physics of Electronic and Atomic Collisions: XVIII International Conference (Aarhus, Denmark, 1993)	93-74103	1-56396-290-X

	Title	L.C. Number	ISBN
No. 296	The Chaos Paradigm: Developments an Applications in Engineering and Science (Mystic, CT 1993)	93-74146	1-56396-254-3
No. 297	Computational Accelerator Physics (Los Alamos, NM 1993)	93-74205	1-56396-222-5
No. 298	Ultrafast Reaction Dynamics and Solvent Effects (Royaumont, France 1993)	93-074354	1-56396-280-2
No. 299	Dense Z-Pinches: Third International Conference (London, 1993)	93-074569	1-56396-297-7
No. 300	Discovery of Weak Neutral Currents: The Weak Interaction Before and After (Santa Monica, CA 1993)	94-70515	1-56396-306-X
No. 301	Eleventh Symposium Space Nuclear Power and Propulsion (3 Vols.) (Albuquerque, NM 1994)	92-75162	1-56396-305-1 (Set) 156396-301-9 (pbk. set)
No. 302	Lepton and Photon Interactions/ XVI International Symposium (Ithaca, NY 1993)	94-70079	1-56396-106-7
No. 303	Slow Positron Beam Techniques for Solids and Surfaces Fifth International Workshop (Jackson Hole, WY 1992)	94-71036	1-56396-267-5
No. 304	The Second Compton Symposium (College Park, MD 1993)	94-70742	1-56396-261-6
No. 305	Stress-Induced Phenomena in Metallization Second International Workshop (Austin, TX 1993)	94-70650	1-56396-251-9
No. 306	12th NREL Photovoltaic Program Review (Denver, CO 1993)	94-70748	1-56396-315-9
No. 307	Gamma-Ray Bursts Second Workshop (Huntsville, AL 1993)	94-71317	1-56396-336-1
No. 308	The Evolution of X-Ray Binaries (College Park, MD 1993)	94-76853	1-56396-329-9
No. 309	High-Pressure Science and Technology—1993 (Colorado Springs, CO 1993)	93-72821	1-56396-219-5 (Set)
No. 310	Analysis of Interplanetary Dust (Houston, TX 1993)	94-71292	1-56396-341-8
No. 311	Physics of High Energy Particles in Toroidal Systems (Irvine, CA 1993)	94-72098	1-56396-364-7

	Title	L.C. Number	ISBN
No. 312	Molecules and Grains in Space (Mont Sainte-Odile, France 1993)	94-72615	1-56396-355-8
No. 313	The Soft X-Ray Cosmos ROSAT Science Symposium (College Park, MD 1993)	94-72499	1-56396-327-2
No. 314	Advances in Plasma Physics Thomas H. Stix Symposium (Princeton, NJ 1992)	94-72721	1-56396-372-8
No. 315	Orbit Correction and Analysis in Circular Accelerators (Upton, NY 1993)	94-72257	1-56396-373-6
No. 316	Thirteenth International Conference on Thermoelectrics (Kansas City, Missouri 1994)	95-75634	1-56396-444-9
No. 317	Fifth Mexican School of Particles and Fields (Guanajuato, Mexico 1992)	94-72720	1-56396-378-7
No. 318	Laser Interaction and Related Plasma Phenomena 11th International Workshop (Monterey, CA 1993)	94-78097	1-56396-324-8
No. 319	Beam Instrumentation Workshop (Santa Fe, NM 1993)	94-78279	1-56396-389-2
No. 320	Basic Space Science (Lagos, Nigeria 1993)	94-79350	1-56396-328-0
No. 321	The First NREL Conference on Thermophotovoltaic Generation of Electricity (Copper Mountain, CO 1994)	94-72792	1-56396-353-1
No. 322	Atomic Processes in Plasmas Ninth APS Topical Conference (San Antonio, TX)	94-72923	1-56396-411-2
No. 323	Atomic Physics 14 Fourteenth International Conference on Atomic Physics (Boulder, CO 1994)	94-73219	1-56396-348-5
No. 324	Twelfth Symposium on Space Nuclear Power and Propulsion (Albuquerque, NM 1995)	94-73603	1-56396-427-9
No. 325	Conference on NASA Centers for Commercial Development of Space (Albuquerque, NM 1995)	94-73604	1-56396-431-7
No. 326	Accelerator Physics at the Superconducting Super Collider (Dallas, TX 1992-1993)	94-73609	1-56396-354-X
No. 327	Nuclei in the Cosmos III Third International Symposium on Nuclear Astrophysics (Assergi, Italy 1994)	95-75492	1-56396-436-8

	Title	L.C. Number	ISBN
No. 328	Spectral Line Shapes, Volume 8 12th ICSLS (Toronto, Canada 1994)	94-74309	1-56396-326-4
No. 329	Resonance Ionization Spectroscopy 1994 Seventh International Symposium (Bernkastel-Kues, Germany 1994)	95-75077	1-56396-437-6
No. 330	E.C.C.C. 1 Computational Chemistry F.E.C.S. Conference (Nancy, France 1994)	95-75843	1-56396-457-0
No. 331	Non-Neutral Plasma Physics II (Berkeley, CA 1994)	95-79630	1-56396-441-4
No. 332	X-Ray Lasers 1994 Fourth International Colloquium (Williamsburg, VA 1994)	95-76067	1-56396-375-2
No. 333	Beam Instrumentation Workshop (Vancouver, B. C., Canada 1994)	95-79635	1-56396-352-3
No. 334	Few-Body Problems in Physics (Williamsburg, VA 1994)	95-76481	1-56396-325-6
No. 335	Advanced Accelerator Concepts (Fontana, WI 1994)	95-78225	1-56396-476-7 (Set) 1-56396-474-0 (Book) 1-56396-475-9 (CD-Rom)
No. 336	Dark Matter (College Park, MD 1994)	95-76538	1-56396-438-4
No. 337	Pulsed RF Sources for Linear Colliders (Montauk, NY 1994)	95-76814	1-56396-408-2
No. 338	Intersections Between Particle and Nuclear Physics 5th Conference (St. Petersburg, FL 1994)	95-77076	1-56396-335-3
No. 339	Polarization Phenomena in Nuclear Physics Eighth International Symposium (Bloomington, IN 1994)	95-77216	1-56396-482-1
No. 340	Strangeness in Hadronic Matter (Tucson, AZ 1995)	95-77477	1-56396-489-9
No. 341	Volatiles in the Earth and Solar System (Pasadena, CA 1994)	95-77911	1-56396-409-0
No. 342	CAM -94 Physics Meeting (Cacun, Mexico 1994)	95-77851	1-56396-491-0
No. 343	High Energy Spin Physics Eleventh International Symposium (Bloomington, IN 1994)	95-78431	1-56396-374-4

	Title	L.C. Number	ISBN
No. 344	Nonlinear Dynamics in Particle Accelerators: Theory and Experiments (Arcidosso, Italy 1994)	95-78135	1-56396-446-5
No. 345	International Conference on Plasma Physics ICPP 1994 (Foz do Iguaçu, Brazil 1994)	95-78438	1-56396-496-1
No. 346	International Conference on Accelerator-Driven Transmutation Technologies and Applications (Las Vegas, NV 1994)	95-78691	1-56396-505-4
No. 347	Atomic Collisions: A Symposium in Honor of Christopher Bottcher (1945-1993) (Oak Ridge, TN 1994)	95-78689	1-56396-322-1
No. 348	Unveiling the Cosmic Infrared Background (College Park, MD, 1995)	95-83477	1-56396-508-9
No. 349	Workshop on the Tau/Charm Factory (Argonne, IL, 1995)	95-81467	1-56396-523-2
No. 350	International Symposium on Vector Boson Self-Interactions (Los Angeles, CA 1995)	95-79865	1-56396-520-8
No. 351	The Physics of Beams Andrew Sessler Symposium (Los Angeles, CA 1993)	95-80479	1-56396-376-0
No. 352	Physics Potential and Development of $\mu^+\mu^-$ Colliders: Second Workshop (Sausalito, CA 1994)	95-81413	1-56396-506-2
No. 353	13th NREL Photovoltaic Program Review (Lakewood, CO 1995)	95-80662	1-56396-510-0
No. 354	Organic Coatings (Paris, France, 1995)	96-83019	1-56396-535-6
No. 355	Eleventh Topical Conference on Radio Frequency Power in Plasmas (Palm Springs, CA 1995)	95-80867	1-56396-536-4
No. 356	The Future of Accelerator Physics (Austin, TX 1994)	96-83292	1-56396-541-0
No. 357	10th Topical Workshop on Proton-Antiproton Collider Physics (Batavia, IL 1995)	95-83078	1-56396-543-7
No. 358	The Second NREL Conference on Thermophotovoltaic Generation of Electricity	95-83335	1-56396-509-7
No. 360	The Physics of Electronic and Atomic Collisions XIX International Conference (Whistler, Canada, 1995)	95-83671	1-56396-440-6

	Title	L.C. Number	ISBN
No. 361	Space Technology and Applications International Forum (Albuquerque, NM 1996)	95-83440	1-56396-568-2
No. 362	Two-Center Effects in Ion-Atom Collisions (Lincoln, NE 1994)	96-83379	1-56396-342-6
No. 363	Phenomena in Ionized Gases XXII ICPIG (Hoboken, NJ, 1995)	96-83294	1-56396-550-X
No. 364	Fast Elementary Processes in Chemical and Biological Systems (Villeneuve d'Ascq, France, 1995)	96-83624	1-56396-564-X
No. 365	Latin-American School of Physics XXX ELAF Group Theory and Its Applications (México City, México, 1995)	96-83489	1-56396-567-4
No. 366	High Velocity Neutron Stars and Gamma-Ray Bursts (La Jolla, CA 1995)	96-84067	1-56396-593-3
No. 367	Micro Bunches Workshop (Upton, NY, 1995)	96-83482	1-56396-555-0
No. 368	Acoustic Particle Velocity Sensors: Design, Performance and Applications (Mystic, CT, 1995)	96-83548	1-56396-549-6
No. 369	Laser Interaction and Related Plasma Phenomena (Osaka, Japan 1995)	96-85009	1-56396-445-7
No. 370	Shock Compression of Condensed Matter-1995 (Seattle, WA 1995)	96-84595	1-56396-566-6
No. 371	Sixth Quantum 1/f Noise and Other Low Frequency Fluctuations in Electronic Devices Symposium (St. Louis, MO, 1994)	96-84200	1-56396-410-4
No. 372	Beam Dynamics and Technology Issues for + - Colliders 9th Advanced ICFA Beam Dynamics Workshop (Montauk, NY, 1995)	96-84189	1-56396-554-2
No. 373	Stress-Induced Phenomena in Metallization (Palo Alto, CA 1995)	96-84949	1-56396-439-2
No. 374	High Energy Solar Physics (Greenbelt, MD 1995)	96-84513	1-56396-542-9
No. 376	Chaos and the Changing Nature of Science and Medicine: An Introduction (Mobile, AL 1995)	96-85220	1-56396-442-2
No. 377	Space Charge Dominated Beams and Applications of High Brightness Beams (Bloomington, IN 1995)	96-85165	1-56396-625-7